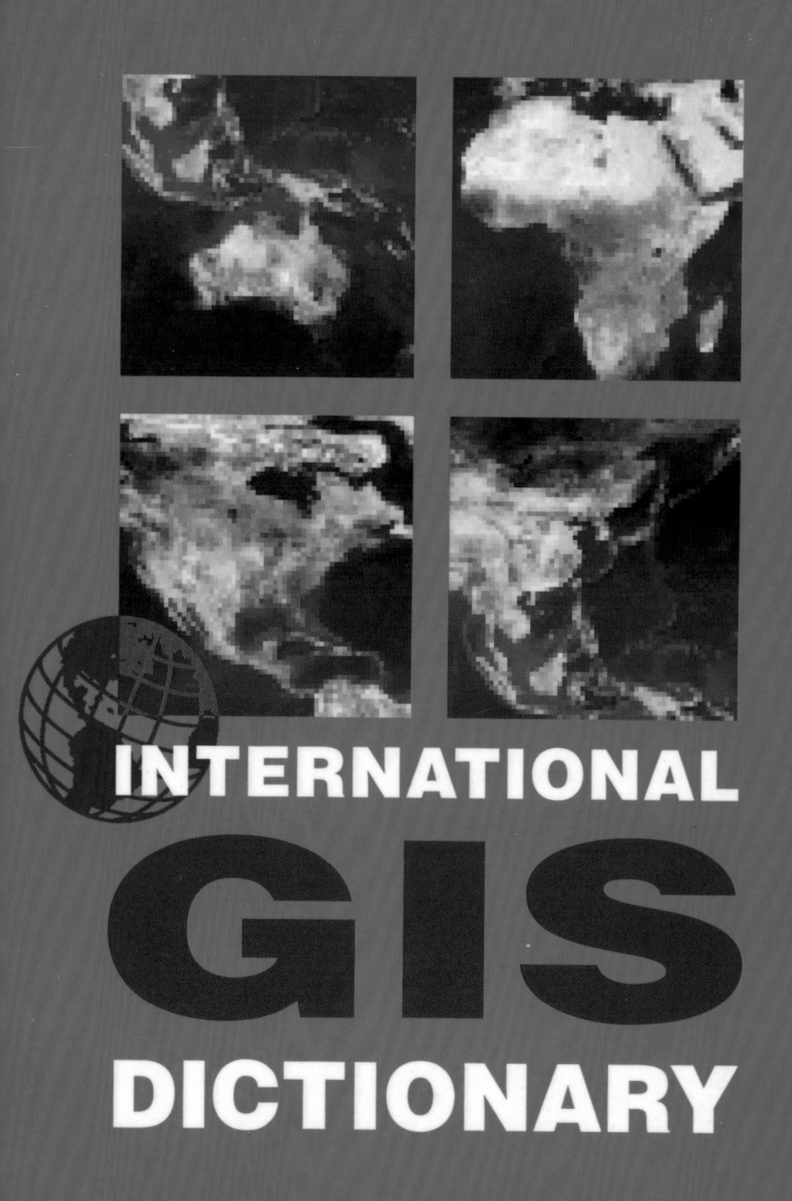

INTERNATIONAL GIS DICTIONARY

Rachael McDonnell & Karen Kemp

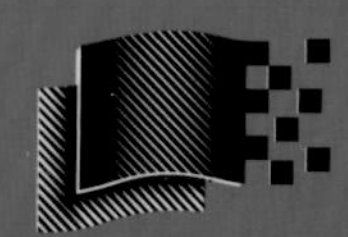

INTERNATIONAL GIS DICTIONARY

Rachael McDonnell & Karen Kemp

GeoInformation International
A division of Pearson Professional Ltd
307 Cambridge Science Park
Milton Road
Cambridge
CB4 4ZD
and Associated Companies throughout the world.

Copublished in the United States with John Wiley & Sons, Inc,
605 Third Avenue, New York, NY 10158

First published 1995

British Library Cataloguing in Publication data
A catalogue entry for this title is available from the British Library.

ISBN 1-899761-19-5

Library of Congress Cataloguing in Publication data
A catalogue entry for this title is available from the Library of Congress.

ISBN 0-470-23607-8 (USA only)

Printed in the United Kingdom
by Bell and Bain, Glasgow

Set in 9/17 Times

Cover photograph: Tony Stone Images (Earth Imaging)

PREFACE

Most people would now agree that GIS can be profitably incorporated into activities encompassing a wide range of different disciplines and application areas. As a result, there is a large and growing group of students and trained professionals who need to master this new field very quickly. However, for these newcomers, and even for many old hands, the literature associated with GIS often seems impenetrable. As in any field, the jargon and acronyms are largely incomprehensible to the uninitiated, and many words that have a familiar interpretation in everyday usage take on a specific meaning in the GIS context. Such an evolving lexicon reflects the dynamism but also the immaturity of this field.

This large and complex vocabulary has developed for a number of reasons. Continuing progress in spatial data theory, as well as in computing technology itself, spawns many new words. Software companies often develop unique vocabularies for their own programs' features and functionality, and users tend to adopt and use these as generic terms. GIS continues to evolve in a variety of different disciplines and organizational backgrounds. Thus, many discipline-specific terms have become more widely used outside these source fields, while at the same time, different words may be used for similar concepts.

The need to establish some common usage of terminology is obvious. Although it is still too early to sanction the definition of many terms, here we have endeavoured to further the understanding of the most common, basic GIS vocabulary. As far as possible, we have avoided including software specific terms. This dictionary has roots in the original AGI dictionary which was first published by the Standards Committee of the AGI in 1991 and was edited by Rob Walker. The *International GIS Dictionary* has been thoroughly updated and revised to include terms from both sides of the Atlantic and from related disciplines which are becoming increasingly important to people using GIS. While it will be many years before a dictionary can encompass the full linguistic richness of the GIS field, we hope that this small contribution will support the work of current GIS professionals and give confidence to students entering the field.

Rachael A McDonnell, OXFORD

Karen K Kemp, SANTA BARBARA

AUGUST 1995

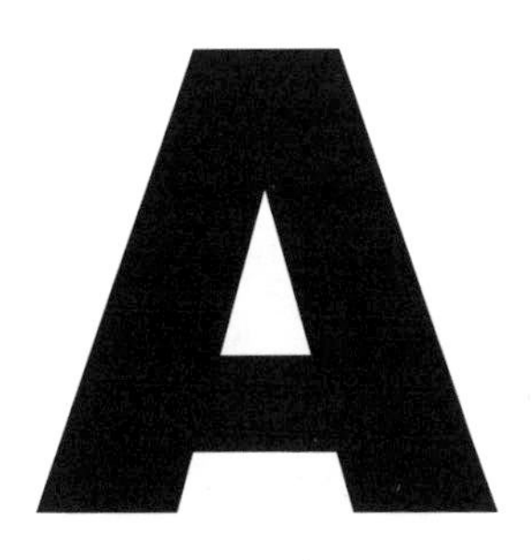

2D

A description of an ENTITY in terms of two spatial dimensions. Locations in these two dimensions are often identified by a pair of co-ordinates, x and y.

2.5D

A representation of an ENTITY in which the third dimension is constrained to a simple relationship with the two horizontal spatial dimensions, such that z is a single valued function of x and y. There can be only one Z-VALUE for each location. This third dimension is often used to represent an ATTRIBUTE VALUE rather than a spatial co-ordinate. Digital representations of elevations are often 2.5D. DTMs are examples of this.

3D

A representation of an ENTITY in terms of three spatial dimensions. Generally these three dimensions are represented as x, y, and z co-ordinates. It is possible to have data about the entity at several identical x,y locations, each with a different Z-VALUE. For example, digital representations of deep-water oceanic currents are often 3D.

AAG

ASSOCIATION OF AMERICAN GEOGRAPHERS A non-profit making organization founded to advance professional studies in geography and to encourage the application of geographic research in education, government, and business. Membership includes professionals in business and government as well as academics and other scholars in the US, Canada and abroad.

ABSOLUTE CO-ORDINATE

The location of a point defined with respect to a specified co-ordinate system.
See RELATIVE CO-ORDINATE.

ABSORPTION

In remote sensing the term refers to the taking up of ELECTROMAGNETIC ENERGY by atmospheric or surface matter. Variation in absorption patterns helps to distinguish between different materials.

ABSTRACTION

A represention of a real-world object. For example, a road may be represented as a CENTRE LINE in one application and as an area bounded by kerblines in another.

ACCURACY

A measure of the numerical difference between observations, computations, or estimates and the true values or the values accepted as being true. Accuracy relates to the correctness of the value result. It is distinguished from precision, which relates to the exactness. In GIS, accuracy may relate to the spatial position of an ENTITY, or to its given ATTRIBUTE VALUE. This accuracy may be expressed in terms of observations or results expected to be correct or which fall within a given margin of ERROR, for example, '20 per cent of the observations are within 10 m of the true elevation'. It may also be expressed in terms of probability, for example, 'each value has a 95 per cent chance of being correct'.

ACROSS-TRACK SCANNING

REMOTE SENSING systems that build up a two-dimensional image of the underlying ground by scanning from side to side and in a direction at right-angles to the aircraft's or satellite's direction of movement. This is achieved through the use of an oscillating mirror which directs the electromagnetic radiance onto the sensing devices. It is also known as whiskbroom scanning. The THEMATIC MAPPER sensor on board the

Landsat series of satellites uses this scanning method.

Across-track scanning.

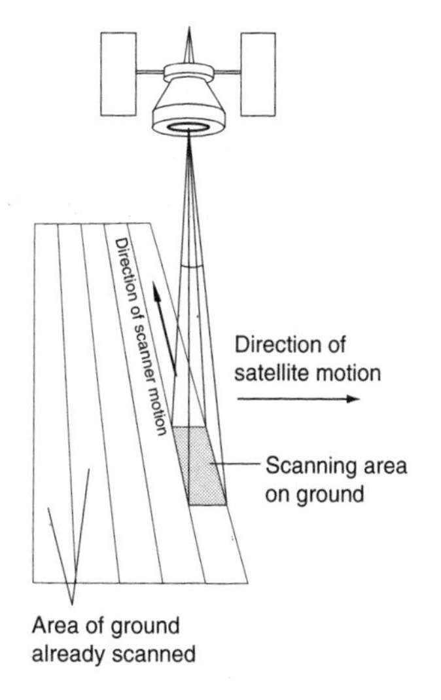

ACSM

AMERICAN CONGRESS ON SURVEYING AND MAPPING A national association for professionals in surveying, cartography, geodesy, GIS, land information systems, and related fields.

ACTIVE SENSOR

A sensor that transmits its own ELECTROMAGNETIC ENERGY to illuminate a scene. Radar is commonly used in remote sensing systems and supplies its own microwave energy. A flash unit linked to a camera is a more familiar example of an active sensor.

ADDRESS

A means of referencing an entity for the purposes of unique identification and location. The postal address is a familiar example where the given reference refers to a distinct domestic or business unit. The term may also be used to refer to a location in computer memory.

ADDRESS MATCHING

1) A procedure which allows two data files to be related through a common address field. If one file has geographic co-ordinates, the data in the other file becomes geocoded. **2)** A procedure by which the items in a non-geocoded database are geocoded to a street network through their addresses. Locations are determined based on address ranges stored for each street segment.

ADJACENCY

The sharing of a common boundary of two regions or polygons.

ADVANCED VERY HIGH RESOLUTION RADIOMETER

See AVHRR.

AFFINE TRANSFORMATION

After DIGITIZING it is necessary to transform the SPATIAL DATA from digitizer co-ordinates to real-world ones prior to any analysis. The affine transformation converts digitizer to CARTESIAN CO-ORDINATES through the following equations:

$x_r = A + Bx_d + Cy_d$

$y_r = D + Ex_d + Fy_d$

where x_r, y_r = Cartesian co-ordinates; x_d, y_d = digitizer co-ordinates and A, B, C, D, E and F are derived co-efficients.

See also HELMERT TRANSFORMATION.

AGGREGATION

The grouping together of a selected set of like entities to form one ENTITY. In some GIS processes, groups of adjacent areal units are linked to form larger units, often as part of a spatial unit hierarchy. For example, UK ELECTORAL WARDS may be grouped into districts. Any attribute data relating to these individual spatial units is also grouped or is summarized to give statistics for the new spatial unit.

See also GENERALIZATION.

AGI

ASSOCIATION FOR GEOGRAPHIC INFORMATION A UK multi-disciplinary non-governmental organization dedicated to the advancement of the use of geographically related information. Its mission is *to spread the benefits of GIS throughout the community and to help all users and vendors of GIS.*

AI

See ARTIFICIAL INTELLIGENCE.

ALGORITHM

A finite, ordered set of well-defined rules for the solution of a problem.

ALIASING

An unwanted visual effect caused by insufficient sampling resolution or inadequate filtering to define the object completely. This is most commonly seen as a jagged edge along the object's boundary, or along a line.

ALONG-TRACK SCANNING

REMOTE SENSING systems that build up a two-dimensional image of the ground by scanning perpendicular to the direction of movement of the aircraft or satellite. This is achieved through the use of a linear array of many detectors which records each scan line at the same time. It is also known as pushbroom scanning. The sensors on board the SPOT satellite series employ this type of sensing.

Along-track scanning.

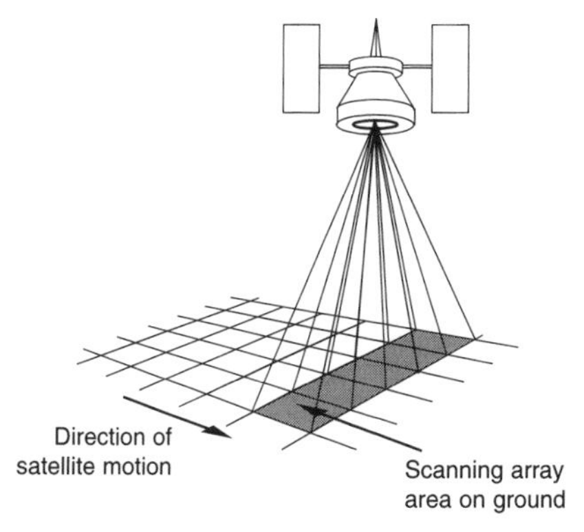

ALPHANUMERIC

A term which describes any letters, numbers, and punctuation marks treated as characters. It is not the same as '*numeric*', which refers to numbers that have a value and can be used in mathematical calculations. The term is also used as an adjective to describe items, such as keyboards, which allow these characters to be entered into a computer in readable code.

AMERICAN CONGRESS ON SURVEYING AND MAPPING

See ACSM.

AMERICAN NATIONAL STANDARDS INSTITUTE

See ANSI.

AMERICAN STANDARD CODE FOR INFORMATION INTERCHANGE

See ASCII.

AM/FM

AUTOMATED MAPPING / FACILITIES MANAGEMENT A term used to describe the collection and management of spatial digital records for use in industries involved in public works or in the management of utilities. These systems combine digital mapping functionality with systems for managing spatial and non-spatial databases describing the infrastructure of the organization.

AM/FM/GIS INTERNATIONAL OF EUROPE

An independent non-profit-making organization which aims to promote, communicate and advise on international activities related to AM/FM and GIS. The organization is managed by a board of directors on which each national section or organization has a seat.

See also AM/FM INTERNATIONAL.

AM/FM INTERNATIONAL

A US-based non-profit-making educational association to help users of AM/FM by promoting information exchange, educational

opportunities, and related research and development. The organization has divisions in various countries and regions.
See also AM/FM/GIS INTERNATIONAL OF EUROPE.

ANALOGUE

A measurement or a representation of an ENTITY which may vary continuously in space or time and which may have a value at any degree of precision. In GIS this concept is used to describe data which is stored or displayed in graphical or pictorial form, as opposed to digital (numeric) form.

ANNOTATION

The ALPHANUMERIC text, labels, scale bar, and title placed on a map to identify entities and to give the user information necessary to interpret it. Street, place, and river names are examples.

ANSI

AMERICAN NATIONAL STANDARDS INSTITUTE A national committee which co-ordinates voluntary standards activities for the US. It has close links with the ISO. For example, it has been responsible for setting standards for digital transfer formats, for DBMS query languages, and for network protocols to allow computers to communicate with each other.

APPLICATION

A practical use of computer software, systems or concepts.

APPLICATIONS PACKAGE

A set of specialized computer programs and associated documentation, usually supplied by an outside agency or software house, for practical usage. Ideally, applications packages allow a non-computer specialist to use the computer without learning complex programming languages. GIS software packages may often be customised for a particular application.

ARC

A line described by an ordered sequence of points. It is a fundamental concept in the VECTOR DATA MODEL. Two or more arcs are joined by a node and several arcs may be linked together in a loop to form an area or polygon.

ARCHIVE

A store of historical records and data.

AREA

1) A fundamental spatial unit consisting of a bounded, continuous 2D entity. Its extent is usually defined in terms of an external polygon or by a contiguous set of grid cells.

2) A mathematical calculation of the size of a 2D entity.

ARRAY

A set of data values held in the form of a line or grid. The ATTRIBUTE VALUES held in the cells or a RASTER DATA MODEL may be considered to be in such a form.

ARTIFICIAL INTELLIGENCE

A means of imitating or formalizing advanced cognitive knowledge into a computer-manageable format which is then used in problem solving or to generate new information. It is often referred to as AI.

ASCII

AMERICAN STANDARD CODE FOR INFORMATION INTERCHANGE A standard coding system used to represent ALPHANUMERIC characters within a computer. As a standard, it enables the transfer of some data between different computer environments by representing them with a common set of symbols.

ASPATIAL DATA

Data associated with spatially referenced elements. The term is often used synonymously with ATTRIBUTE VALUE.

ASPECT

The direction in which a topographic slope faces, usually expressed in terms of degrees from the north. Many GIS provide functions which generate aspect from continuous elevation surfaces.

ASSOCIATION FOR GEOGRAPHIC INFORMATION

See AGI.

ASSOCIATION OF AMERICAN GEOGRAPHERS

See AAG.

ASSOCIATION OF CHINESE PROFESSIONALS IN GEOGRAPHIC INFORMATION SYSTEMS–ABROAD

See CPGIS.

ATMOSPHERIC ABSORPTION

This is the loss of ELECTROMAGNETIC ENERGY through particles, such as ozone and water vapour, in the atmosphere. The extent of the loss is controlled by the size of the particles and the wavelength of the energy.

ATMOSPHERIC WINDOW

The wavebands at which the atmosphere readily transmits ELECTROMAGNETIC ENERGY without attenuation. It is within these wavebands that remote sensing detectors are able to acquire data about the Earth's surface.

ATTRIBUTE VALUE

A value or property that is a characteristic of a spatial element. For example, a contour may be represented in a GIS as an arc with an elevation value associated with it, where elevation is an attribute value. A grid cell, a polygon, or an object may be described in terms of a particular soil type, a population density, or a plant association, all of which are attributes.

AURISA

AUSTRALASIAN URBAN AND REGIONAL INFORMATION SYSTEMS ASSOCIATION

A multi-disciplinary organization whose members share an interest in GIS. It aims to promote the development of urban and regional information systems, to address related public policy issues, and to provide a forum for the exchange of information in the region.

See also URISA.

AUSTRALASIAN URBAN AND REGIONAL INFORMATION SYSTEMS ASSOCIATION

See AURISA.

AUTOCORRELATION

Generally, a statistical measure which describes the extent to which one property changes as another changes. In GIS it is often used to refer to a statistical measure which describes the extent to which the value of an attribute at geographically referenced points changes as a function of the distance and orientation between them.

AUTOMATED CARTOGRAPHY

The process of producing maps with the aid of computer software. Spatial data processing is limited to that required to support map production.

AUTOMATED DIGITIZING

This refers to methods involved in the conversion of analogue data to digital form which involves little or no operator intervention during the data capture stage. It usually requires considerable reconstruction of geometry and TOPOLOGY to recover the structure and meaning of the information after the initial encoding of it. A scanner coupled with automatic RASTER-TO-VECTOR CONVERSION procedures is a good example of this type of data capture.

AUTOMATED FEATURE RECOGNITION

The identification of spatial patterns within a spatial data set using computer algorithms. These techniques are frequently used to delineate landforms from data in a DTM or in creating linear representations of roads from scanned maps.

AUTOMATED MAPPING

See AUTOMATED CARTOGRAPHY.

AUTOMATED MAPPING AND FACILITIES MANAGEMENT

See AM/FM.

AUTOMATED NAME PLACEMENT

The addition of LABELS to spatial features on a map by a computer. The positioning of the label requires an unambiguous association to be achieved between it and the corresponding feature. More sophisticated programs enable overlapping names to be intelligently shifted.

AVHRR

ADVANCED VERY HIGH RESOLUTION RADIOMETER An ACROSS-TRACK SCANNING remote sensing instrument on board the NOAA series of satellites. It senses in the red and near-, mid-, and thermal-infrared wavebands of the ELECTROMAGNETIC SPECTRUM. Its low SPATIAL RESOLUTION gives images at the continental scale.

AZIMUTHAL PROJECTION

A type of map projection constructed as if a planar surface were placed at a tangent to the Earth's surface and the area to be mapped were projected onto it. Each map has only one standard point, the centre. Also known as a zenithal projection.

Schematic representation of the Azimuthal projection.

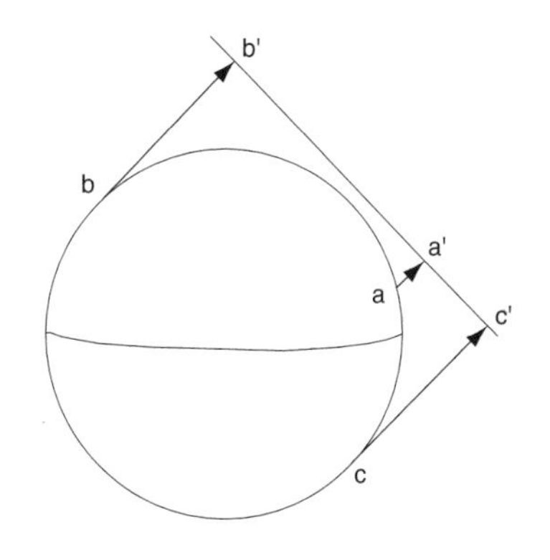

BACKSCATTER

The dispersion of ELECTROMAGNETIC ENERGY by small particles back towards its source.

BAND RATIOS

An IMAGE PROCESSING technique which is used to enhance the contrast between features in remotely sensed images using two or more wavebands. The ratio is derived by dividing the DIGITAL NUMBERS held for one band by that held for the other for exactly the same cell in the images. This technique is particularly useful in eliminating differences in a scene that arise from variations in illumination alone.

BASE MAP

1) A set of topographical data displayed in map form, providing a reference for user's data.

2) A BINARY digital map in a GIS which defines the area within which analysis of other SPATIAL DATA is undertaken.

BASIC LAND AND PROPERTY UNIT

See BLPU.

BASIC SPATIAL UNIT

A fundamental areal unit which has homogeneous properties in the context of a particular subject such as administrative responsibility or ownership. It is often referred to as a BSU.

BATHYMETRIC CHART

A map showing the depths of a body of water.

BAUD RATE

A measure which describes the speed of the transmission of single digital elements over a communications line. This might be between devices such as modems, or between a computer and a printer.

BENCHMARK

1) A test devised in terms of a user's requirements to enable comparisons to be made between different computer software or systems.

2) A surveying term for a mark whose height relative to a DATUM is known.

BEST LINEAR UNBIASED ESTIMATE

BLUE The result of an interpolation function, which was optimized with chosen interpolation weights, at a given variable point.

BILINEAR INTERPOLATION

A mathematical technique to estimate a grid cell value based on data held in the neighbouring four cells. It is often used in the resampling of a raster data set to create a new one with a different cell size or internal geometry. It is a standard technique in remote sensing IMAGE PROCESSING.

Bilinear interpolation for cell x based on data held in the four neighbouring cells y.

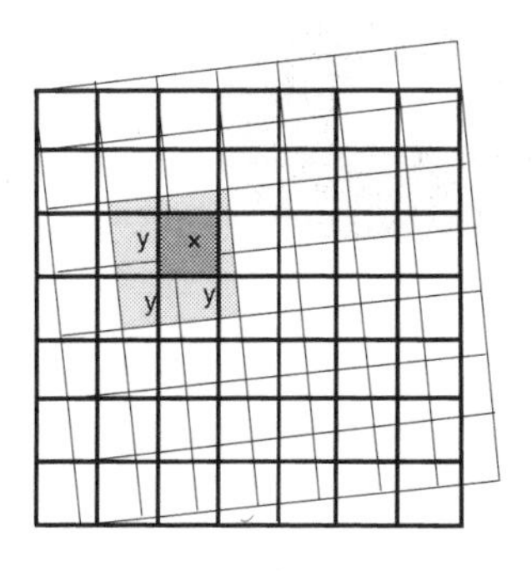

BINARY

A base 2 number that uses only the data values of 0 or 1. It is the fundamental basis of all digital computing.

BINARY DIGITAL MASK

An analytical technique which is used to create a BINARY IMAGE of two classes based on a numerical threshold. For example, an image may be reclassified according to whether an area is above or below an elevation.

BINARY IMAGE

A raster image whose cells contain only one of two possible values, for example, 1 or 0. Binary images result from logical expressions and other functions.

BINARY LARGE OBJECT

See BLOB.

BIT

An abbreviated term for binary digit, and the smallest unit of computer data.

BITMAP

An unstructured grid image in which data values are recorded as either 1 or 0. The term is often used to refer to graphics composed of pixels, for example, scanned maps or photographs.

BIVARIATE INTERPOLATION

An INTERPOLATION technique which uses two variables to specify an equation for a third. Often used to interpolate elevation data.

BLOB

BINARY LARGE OBJECT An area within a raster data set that may be considered as a contiguous feature.

BLOCK ENCODING

A data compression technique used to reduce the required storage space for raster data.

BLOCK KRIGING

A PIECEWISE form of KRIGING based on grid cells.

BLPU

BASIC LAND AND PROPERTY UNIT The physical extent of a contiguous area of land under uniform property rights.

BLUE

See BEST LINEAR UNBIASED ESTIMATE.

BOOLEAN OPERATORS

Logical operators which allow combinations of sets. They include the AND, OR, NOT, and XOR (exclusive OR) terms.

BOUNDARY

A continuous line which delineates the edge of a polygon or study area.

B-SPLINE

A set of splines that, fitted together sequentially, are used to approximate and smooth a line joining a set of points. It is used for smoothing digitized lines, such as the boundaries on geomorphological or ecological maps, to improve their appearance in graphic display.

BSU

See BASIC SPATIAL UNIT.

BUFFER

1) A region of a specified width around a point, line, or area. It is a type of proximity analysis that is supported by most GIS and is defined in such systems by real-world distances from one or more map elements.
2) A temporary storage area held in computer memory or on disk.

BYTE

A unit of computer storage of BINARY data usually comprising eight BITS, and equivalent to a character. Computer memory and storage is measured in this unit, giving rise to terms such as Kilobyte (approximately one thousand bytes), Megabyte (approximately one million bytes) and Gigabyte (approximately one thousand million bytes).

CAD

COMPUTER-AIDED DESIGN / DRAFTING / DRAWING A computer-based information processing system which supports engineering planning and illustrating activities. Many such systems provide advanced features such as solid modelling. The TOPOLOGY of entities is not preserved in the database.

CADASTRAL SURVEY

A survey based on the precise measurement and marking of land parcel boundaries, and thus concerned with property ownership and often with value.

CADASTRE

A public register of the value and ownership of the land of a country, state, or municipality.

CAG

CANADIAN ASSOCIATION OF GEOGRAPHERS An association which aims to promote geographic research and teaching, and which represents the profession in the scientific and business communities.

CANADIAN ASSOCIATION OF GEOGRAPHERS

See CAG.

CANADIAN INSTITUTE OF GEOMATICS

See CIG.

CARTESIAN CO-ORDINATES

Co-ordinates locating points in space expressed by reference to two or three perpendicular axes.

A 2D Cartesian co-ordinate whose position is defined in terms of its distance from an origin along two perpendicular axes.

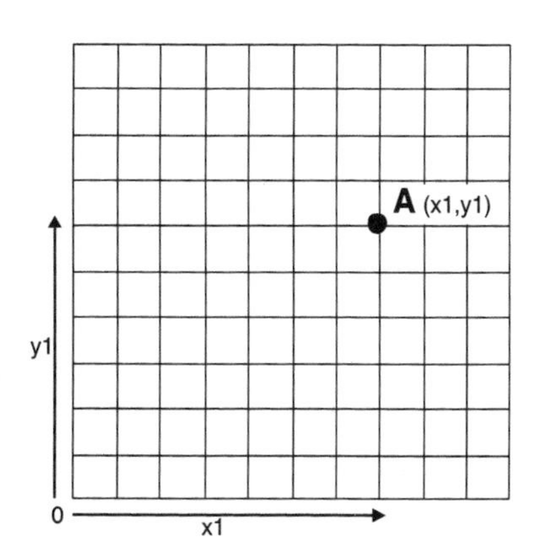

CARTOGRAM

A map in which the areas or distances have been distorted to show variations in magnitude of themes such as commuting times or population density.

CARTOGRAPHY

The art and science of the organization and communication of geographically related information into maps or charts. It can include all stages of their construction, from data acquisition to presentation and use.

CCD

See CHARGE-COUPLED DEVICE.

CCITT

COMITÉ CONSULTATIF INTERNATIONAL DE TÉLÉGRAPHIQUE ET TÉLÉPHONIQUE

An international body primarily addressing telecommunications standards.

CCT

See COMPUTER-COMPATIBLE TAPE.

CELL

The basic spatial element of spatial information in the RASTER DATA MODEL description of spatial ENTITIES. It is usually square or rectangular in shape, although hexagonal and circular areas have also been used.

CELLULAR AUTOMATA

A form of computer modelling that results in simple holistic structures constructed from simple transformation rules applied iteratively to individual grid cell data.

CEN

COMITÉ EUROPÉEN DE NORMALISATION

The regional standards group for Europe. It is not a recognized standards development organization, and so cannot contribute directly to ISO. It functions broadly as a European equivalent to ISO and its key goal is to harmonize standards produced by the standards bodies of its member countries. Membership is open to countries of the European Union and the European Free Trade Association.

CENSUS TRACT

A small, locally delineated statistical area used by the US census. Census tracts generally have stable boundaries and, when first established, are designed to have relatively homogeneous demographic characteristics and an average population of 4 000. There are over 50 000 census tracts covering the US.

See also ENUMERATION DISTRICT.

CENTRE LINE

A line digitized along the centre of a linear feature. Rivers and roads are often represented in this way.

CENTRE NATIONAL D'INFORMATION GÉOGRAPHIQUE

See CNIG.

CENTROID

The position at the centre of a one- or two-dimensional (2D) ENTITY. Calculating the centre of an irregularly shaped polygon requires the use of geometrical algorithms. Centroids are often used as reference points for polygons in a GIS.

CENTRO NACIONAL DE INFORMAÇÃO GEOGRÁFICA

See CNIG.

CERCO

COMITÉ EUROPÉEN DES RESPONSABLES DE LA CARTOGRAPHIE OFFICIELLE The European committee of directors of official national mapping agencies (such as the UK's Ordnance Survey and France's Institut Géographique National) which functions under the auspices of the Council of Europe.

CGM

COMPUTER GRAPHICS METAFILE

A standard file format specification (ISO 8632) for the storage and transfer of graphic information.

CHAIN

A directed sequence of co-ordinates defining a complex line or boundary.

See also LINE.

CHAIN CODE

A data compression technique used with raster data to reduce the required storage space.

CHARGE-COUPLED DEVICE

A micro-electronic memory device made from silicon that stores patterns of electronic charges in a sequential fashion for an array of energy sensitive elements. These elements sense ELECTROMAGNETIC ENERGY and return a voltage measurement according to the intensity of the received input. They are used in cameras and other remote sensing instruments. Also known as CCD.

CHECK PLOT

A graphic output of spatial data which is used to verify either the content or positional accuracy of the information. Accuracy may be checked by direct superimposition of the check plot onto the graphic original used to create the digital record.

CHORLEY REPORT

The common name of the Report of the Committee of Inquiry into the Handling of Geographic Information, completed in the UK in 1987 by a committee chaired by Lord Chorley. The publication of this report and the activity associated with it are recognized as important milestones in the development of GIS.

CHOROPLETH MAP

A thematic map in which quantitative SPATIAL DATA is depicted through the use of shading or colour variations of individual unit areas.

A chloropleth map of the African subcontinent.

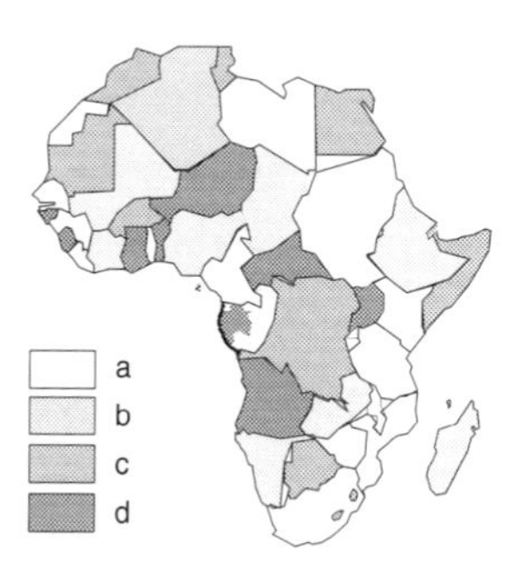

CIG

CANADIAN INSTITUTE OF GEOMATICS

A national association which aims to advance the development of geomatics science in Canada. Formerly the Canadian Institute of Surveying and Mapping (CISM), the organization includes surveyors, hydrographers, photogrammetrists, remote-sensing specialists, GIS/LIS specialists, cartographers, vendors, suppliers, and others in related fields.

CISM

See CIG.

CLASS

A set of ENTITIES possessing certain common ATTRIBUTE VALUES.

CLASSIFICATION

The grouping of ENTITIES into a set of CLASSES according to certain common ATTRIBUTE VALUES.

CLIPPING

The action of trimming a set of SPATIAL DATA by removing all the display elements that lie outside a boundary. This is often done by overlaying a map of the boundary area onto the one which is to be truncated. Clipping is often used for display purposes, eliminating extraneous detail in a graphic.

CNIG

Either the **CENTRO NACIONAL DE INFORMAÇÃO GEOGRÁFICA** (Portugal) or the **CENTRE NATIONAL D'INFORMATION GÉOGRAPHIQUE** (France). Both are national centres for geographic information.

COGO

CO-ORDINATE GEOMETRY A set of algorithms for converting survey data (bearings, distances and angles) into co-ordinate data.

CO-KRIGING

A multi-variate approach to INTERPOLATION using KRIGING, where two or more attributes are used to determine values for a different attribute data set.

COMITÉ CONSULATATIF INTERNATIONAL DE TÉLÉGRAPHIQUE ET TÉLÉPHONIQUE

See CCIT.

COMITÉ EUROPÉEN DE NORMALISATION

See CEN.

COMITÉ EUROPÉEN DES RESPONSABLES DE LA CARTOGRAPHIE OFFICIELLE

See CERCO.

COMPACTION

See COMPRESSION.

COMPLEX DIELECTRIC CONSTANT

A measure of the free electrical charge of a material. The response of substances to radar remote sensing is affected by their electical conductivity with high constants (indicating many free electrical charges) giving a high BACKSCATTER.

COMPOSITE MAP

A single map created by joining together several others.

COMPRESSION

A reduction of file size for data handling and storage. Examples of such methods include QUADTREES and RUN-LENGTH ENCODING. Also known as compaction.

COMPUTER-AIDED DESIGN / DRAFTING / DRAWING

See CAD.

COMPUTER-COMPATIBLE TAPE

CCT Electronic tape media used to transfer digital data from one processing system to another.

COMPUTER GRAPHICS METAFILE

See CGM.

CONNECTIVITY

A term to describe the linking of points or polygons to each other. Many GIS include functions which can be used to determine which spatial ENTITIES are connected to which others.

A record of the connectivity of the point data shown in the illustration would record that node c is connected to nodes b, d, and f, but not to nodes a, e, g, or h.

CONTIGUITY

A term describing the touching of spatial ENTITIES, usually polygons. Many GIS permit analysis that determines which entities are touching a particular spatial unit.

A record of the contiguity of the polygon data shown in the illustration would record that polygon S touches polygons R, Q, and T but not polygon P.

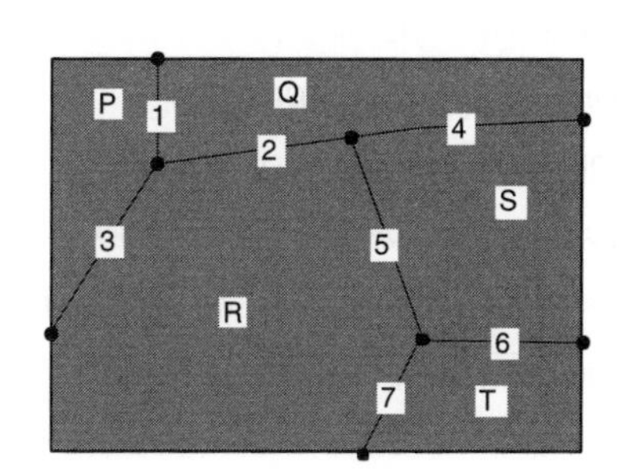

CONTIGUOUS

Similar to ADJACENCY, though often used to refer to a set of spatial ENTITIES, usually polygons, which share common boundaries such that a continuous region with a single, continuous boundary can be identified.

CONTINUOUS DATA

Data which represents a phenomenon that occurs everywhere, that varies continuously through space and that, potentially, could be measured at any location. Elevation, temperature, and air pressure data are good examples of continuous data. In the digital computer, continuous data must be stored using discrete methods such as contours or grids of ATTRIBUTE VALUES at specific points.

CONTOUR

A line connecting a set of points, all of which have the same value. The term is often used in the context of elevation data.

CONTRAST STRETCHING

An IMAGE ENHANCEMENT technique used to accentuate the differences between features within a scene by using the full range of display colours available. The range of DIGITAL NUMBER values in a given scene are transformed to expand across the range of the display values. This ensures that subtle variations in the values are now shown more clearly.

Contrast stretching technique which takes the range of digital number values of the image (in the example, 50–225) and stretches them across the available display level values (in the example, 0–255) to increase the level of distinction within the scene.

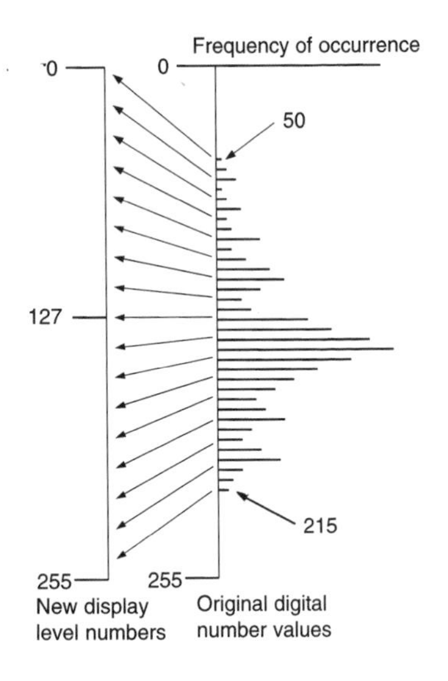

CONTROL

In mapping, this is a system of points with established horizontal and vertical positions which are used as fixed references to which mapped features can be related. The establishment of controls is one of the first steps involved in DIGITIZING.

CO-ORDINATE

x, y and possibly z-values defining a position in terms of a spatial reference framework.

CO-ORDINATE GEOMETRY

See COGO.

CO-ORDINATE POINT

A location in space defined by measured linear or angular positions within a SPATIAL REFERENCE system. In a 2D Cartesian CO-ORDINATE SYSTEM this location is often designated as (x,y).

CO-ORDINATE SYSTEM

A fixed reference framework superimposed onto the surface of an area to designate the position of a point within it. The CARTESIAN CO-ORDINATE system and the system of latitude and longitude used on the Earth's surface are common examples.

CO-ORDINATE TRANSFORMATION

A mathematical process in which a set of co-ordinates representing locations in space is converted to another co-ordinate system.

A co-ordinate transformation.

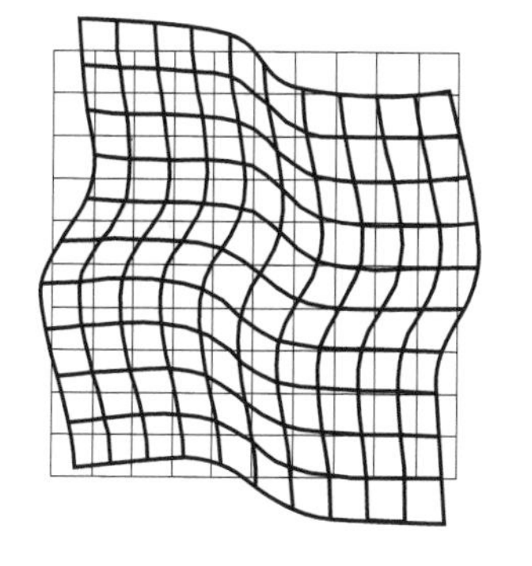

CORNER REFLECTION

The high reflection of ELECTROMAGNETIC ENERGY back to a sensor resulting from the geometry and nature of a surface object.

The corner reflection effect resulting from the geometry of the building.

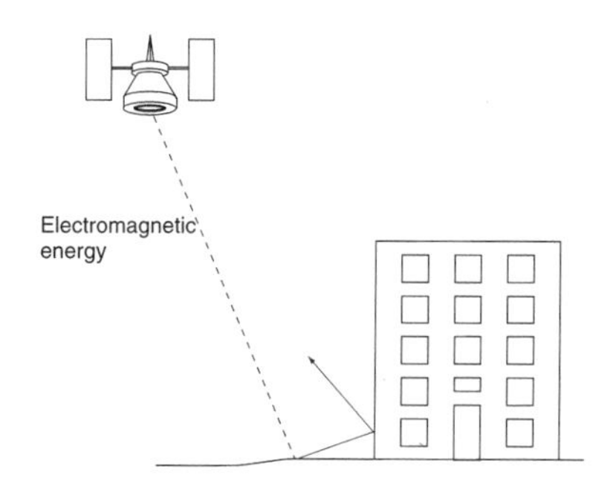

COVERAGE

A term sometimes used to describe a spatial LAYER of information held within a GIS.

CPGIS

ASSOCIATION OF CHINESE PROFESSIONALS IN GEOGRAPHIC INFORMATION SYSTEMS—ABROAD

A non-profit-making organization whose members include students and professionals in GIS, remote sensing, GPS, and related fields. Its aims include the promotion of professional development among members, and the exchange of ideas and knowledge between Chinese GIS professionals in China and abroad, and between Chinese GIS professionals and those of other nationalities.

CUBIC INTERPOLATION

A technique which determines new values for a cell within a grid system based on fitting a third order polynomial surface to a 3 by 3 cell neighbourhood area.

Cubic interpolation for cell x based on data held in the 16 neighbouring cells y.

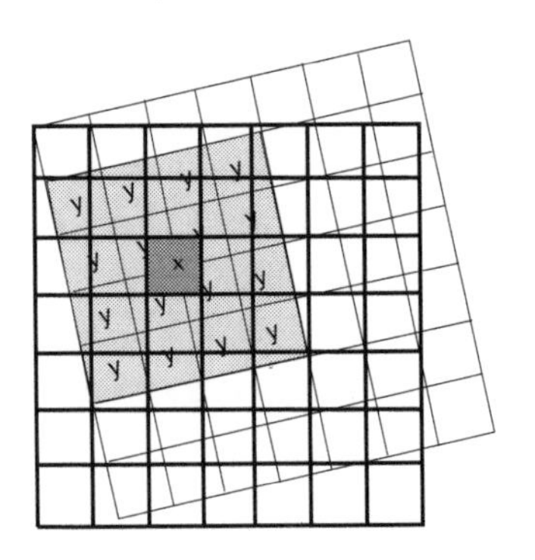

CURSOR

1) A pointer or other symbol on a screen that indicates the active position. Movement of the cursor is generally accomplished through the use of a mouse or other pointing device. **2)** In DIGITIZING, the term cursor refers to the hand-held device used for selecting menu items or for accurately identifying the location of mapped graphic objects during digitizing. This device is also termed a PUCK.

CUSTOMIZATION

A procedure which produces an application or company specific interface and/or database design using a generic software product. For example, a customized version of a commercial GIS product may include menus which allow administration staff who do not know how to use the GIS to enter customer data into the database.

CUT-AND-FILL

This term refers to volumetric calculations that may be made from a DTM for engineering applications such as road design.

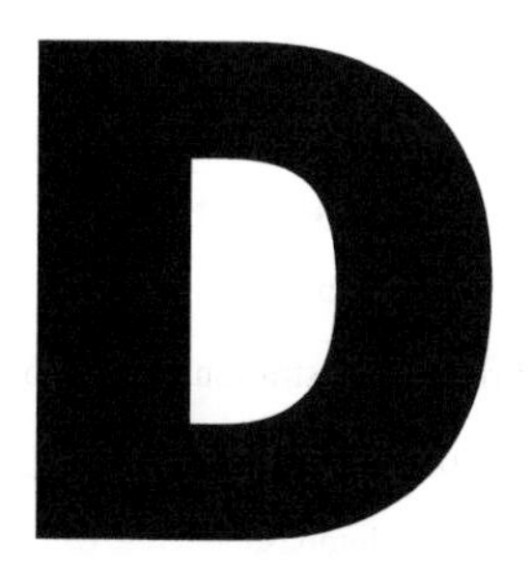

DANGLE

See OVERSHOOT.

DATA

A collection of facts, concepts or instructions in a formalized manner suitable for communication or processing by human beings or by automatic means. In GIS applications they are often observations or measurements of the natural or human environment.

DATABASE

A collection of data organized according to a conceptual schema with a set of procedures for adding, changing, or retrieving data held in this structure.

DATABASE MANAGEMENT SYSTEM

DBMS A collection of software for organizing the information in a database. Typically it contains routines for data input, verification, storage, retrieval, and combination.

DATA CAPTURE

The process of converting data into digital form by DIGITIZING, scanning, or direct recording of real-world phenomena.

DATA CLASSIFICATION

The division of a set of data or ENTITIES into subsets such that entities within each subset share some common features.

DATA COLLECTION

The process of collecting data by survey for conversion to digital form. This includes aerial and satellite surveys, ground surveys, and census surveys.

DATA CONVERSION

The conversion of data from an original form (paper or electronic) into a form suitable for a particular use in a digital form compatible with the computer system, software, and other data being used.

DATA FORMAT

A specification which defines how data is structured in a file and which determines such aspects as the sequence of data items, their length, and necessary HEADER data.

DATA INPUT

The entry of information into a computer. This may be accomplished through the use of a keyboard, DIGITIZER, SCANNER, or from already existing data sets.

DATA ITEM

A sequence of related characters which can be defined as the smallest logical unit of data that can be independently and meaningfully processed. For example, this might be an x,y CO-ORDINATE pair of values.

DATA MODEL

In a general sense it is an ABSTRACTION of the real world which incorporates only those properties thought to be relevant to the application at hand. It would normally define specific groups of ENTITIES, their ATTRIBUTE VALUES, and the relationships between these. In GIS usage it often is used to refer to the mechanistic representation and organization of SPATIAL DATA, common models being the VECTOR DATA MODEL or a RASTER DATA MODEL. It is independent of a computer system and its associated data structures.

DATA QUALITY

An assessment of the completeness, currency, logical consistency, and accuracy of data for a particular purpose.

DATA SET

An organized collection of data with a common theme.

DATA STRUCTURE

A logical arrangement of data used by a specific computer system for data management, storage, and retrieval. It includes a reference linkage system between data items.

DATA TRANSFER

The movement of data from one computer system or software to another. It often requires a change of DATA FORMAT.

DATUM

1) Any point, line, or surface used as a reference for a measurement of another quantity.
2) A model of the Earth used for geodetic calculations.

DBMS

See DATABASE MANAGEMENT SYSTEM.

DELAUNAY TRIANGULATION

A method of fitting triangles to a set of points. The triangles are defined by the condition that the circumscribing circle of any triangle does not contain any other points of the data except the three defining it. It is a method which produces triangles with a low variance in edge length. The resulting triangles may be used as an irregular tessellation for interpolation of other points on a surface.

Delaunay triangulation.

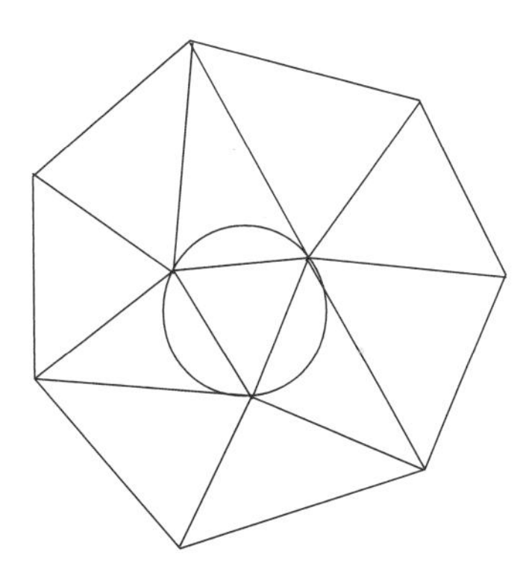

DEM

DIGITAL ELEVATION MODEL A digital representation of the elevation of locations on the land surface. A DEM is often used in reference to a set of elevation values representing the elevations at points in a rectangular grid on the Earth's surface. Some definitions expand DEM to include any digital representation of the land surface, including TINS or digital CONTOURS.
See also DTM.

DENSITY SLICING

An IMAGE ENHANCEMENT technique. The DIGITAL NUMBERS of a remotely sensed image are divided into classes and a display colour is assigned to each class. Sharp contrasts between the different groups are therefore clearly displayed.

DERIVED MAP

A map produced by analysing, altering, or combining other maps rather than from an original survey.

DEVICE CO-ORDINATE

A point reference specified according to a device's own referencing system. The reference system of a DIGITIZER is an example.

DEVICE SPACE

The area defined by the complete set of addressable points of a display device.

DIALOGUE BOX

A pop-up window within a computer program into which data or commands may be entered.

DIFFUSE REFLECTION

The reflection by an object of ELECTROMAGNETIC ENERGY uniformly in all directions. Also known as Lambertian reflection.

Diffuse reflection of electromagnetic energy by the ground surface.

Incoming electromagnetic radiation

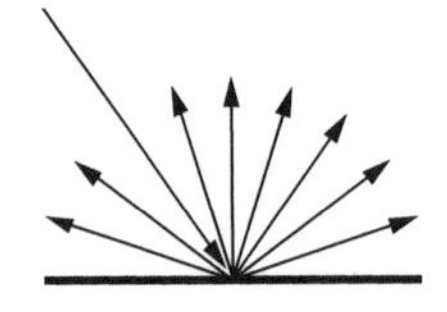

DIGEST

The Digital Geographic Information Working Group (DGIWG) Exchange Standard. A NATO standard for the exchange of geographic data in digital form between defence agencies.

DIGITAL ELEVATION MODEL

See DEM and DTM.

DIGITAL EXCHANGE FORMAT

See DXF.

DIGITAL IMAGE PROCESSING

See IMAGE PROCESSING.

DIGITAL LINE GRAPH

See DLG.

DIGITAL MAPPING

The process of storing and displaying map data in computer-compatible form.

DIGITAL NUMBER

DN In remote sensing, this is a value, usually ranging on a scale between 0 and 255, assigned to the average measured radiance observed by a sensor to indicate its relative intensity.

DIGITAL TERRAIN MODEL

See DTM and DEM.

DIGITIZER

A table or tablet capable of digitally recording the relative position of a CURSOR which an operator moves across the surface.

DIGITIZER CO-ORDINATE

See DEVICE CO-ORDINATE.

DIGITIZING

The conversion of analogue maps and other graphic data into a computer-readable form using a DIGITIZER. There are different methods, for example, AUTOMATED DIGITIZING, POINT DIGITIZING or STREAM DIGITIZING.

DIGITIZING TABLE

See DIGITIZER.

DIME

DUAL INDEPENDENT MAP ENCODING

A spatial DATA STRUCTURE in which line segments are defined both by the nodes which indicate their end points and by the areas which bound each line segment. DIME was the 1970s precursor to the TIGER system now used for the US decennial census and has been subsumed within it.

DIRECTED LINK

A link between two nodes with one direction specified.

DIRICHLET TESSELLATION

See THIESSEN POLYGONS.

DISCRETIZATION

The process of dividing a group of ENTITIES within an area into distinct units. This is particularly important when working with continuous phenomena which cannot be represented directly within the computer.

DISSOLVING

The process of merging two adjacent areas or polygons and the removal of any boundary lines between them.

DISTRIBUTED SYSTEM

A complex computer system where the workload is spread between two or more computers linked together by a communications network.

DLG

DIGITAL LINE GRAPH A US Geological Survey digital map format used to distribute topographical maps in vector form. The digital files contain lists of point co-ordinates describing linear map features.

DN

See DIGITAL NUMBER.

DOTS PER INCH

DPI A unit of measurement expressing the resolution of scanners, monitors, or printing devices. The greater the number of dots per inch, the more detail may be captured or displayed by the device.

DPI

See DOTS PER INCH.

DRAPING

The overlaying of a spatial data set onto a 2.5D elevation surface.

DRUM SCANNER

A SCANNER on which an original document is wrapped around a cylinder which rotates beneath the scanning elements.

DTM

DIGITAL TERRAIN MODEL A digital representation of the elevation of the ground surface generated from elevation as well as topological data. Often used synonymously with DEM, though also used as the more generic term.

See also DEM.

DUAL INDEPENDENT MAP ENCODING

See DIME.

DX 90

A data format for digital hydrographic data, developed by the International Hydrographic Organization (IHO). Together with the IHO's feature coding scheme and a number of digitizing conventions, it comprises the IHO's Transfer Standard for Digital Hydrographic Data.

DXF

DIGITAL EXCHANGE FORMAT A format for transferring digital data between computer systems, widely used as a de facto standard in the engineering and construction industries.

DYNAMIC SEGMENTATION

The process by which the location of points along linear features of a network is computed as needed rather than being stored at the time the database is created.

EDGE

A line between two nodes or points bounding one or more faces of a spatial ENTITY.

EDGE ENHANCEMENT

An IMAGE ENHANCEMENT technique used to accentuate local contrasts in tone usually in a certain direction. It enhances differences in radiance values by passing a FILTER over a gridded data set, usually a remotely sensed image.

EDGE MATCHING

The process of comparing and adjusting data along adjacent edges of digital map sheets or some other unit of storage, to ensure they agree in both positional and attribute terms.

EDI

ELECTRONIC DATA INTERCHANGE The exchange of processable data electronically between computers.

EDIGeO

A data transfer format strongly based on DIGEST, adopted as a French experimental standard Z-13-150.

EDIT

The process of adding, deleting, and changing data.

ELECTORAL DISTRICT

The fundamental spatial unit used in the division of voting areas in the US and Canada.

ELECTORAL WARD

The fundamental spatial unit used in the division of voting areas in the UK.

ELECTROMAGNETIC ENERGY

A form of radiant energy that is transferred by radiation. It travels in a sinusoidal, harmonic wave form and it is measured in terms of its wavelength and frequency. There are many common forms of this energy including heat, visible light, and microwaves. Its detection by remote sensing systems provides an important source of data for many fields of study.

Electromagnetic energy.

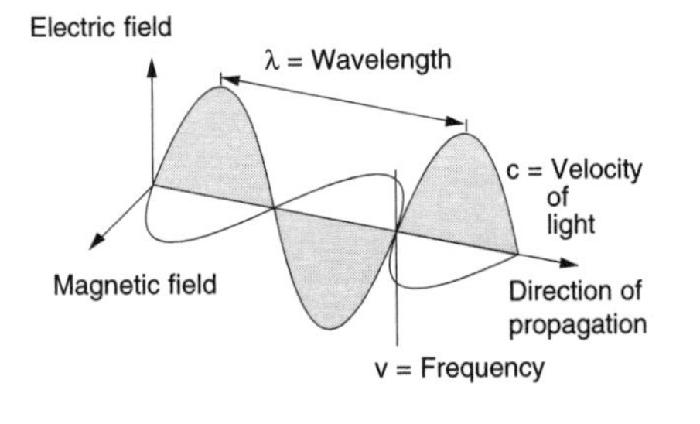

ELECTROMAGNETIC SPECTRUM

The continuum of wavelengths of ELECTROMAGNETIC ENERGY. It is often divided into nominal classes along the spectrum, for example, x-rays, ultraviolet, the visible, thermal, infrared, and radio waves.

The electromagnetic spectrum.

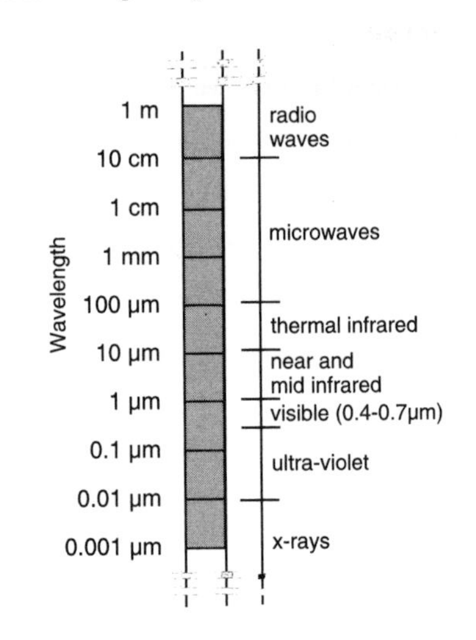

ELECTRONIC DATA INTERCHANGE

See EDI.

ELECTROSTATIC PLOTTER

A device for plotting or printing images by placing small electrical charges on the paper so that a dark or coloured powder (toner) wil adhere to them.

EMULATOR

A software package or program which imitates the functions of a hardware device o other software.

ENCODING

The assignment of a unique code to each unit of data, for example, the encoding of English-language characters using the ASCII codes. The term is also used in reference to the conversion of data to a digital format.

ENDLAP

Overlapping of the coverage of aerial photographs in the direction of the flight. This allows stereoscopic viewing of the scene and ensures all the terrain is photographed in that direction.

Endlap of two aerial photographs.

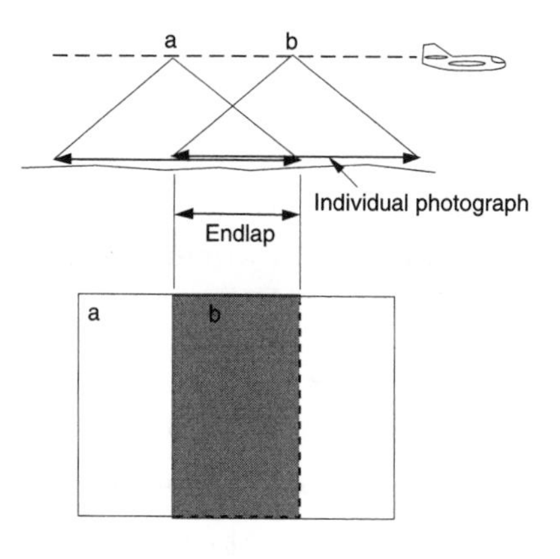

ENTITY

A general term for a real-world thing or digital phenomenon. A house, a road segment, and an ELECTORAL WARD are all examples.

ENTITY CLASS

A specified group of real-world things, for example, a road network.

ENTITY RELATIONSHIP MODEL

A logical way of describing real-world things and their relationships – a necessary first step during database design. This leads to the development of a conceptual model of the relationships between the elements which will be contained within the database and ensures an efficient database design. Also know an an ER model.

ENUMERATION DISTRICT

The basic area unit, containing approximately 150 households, used by the UK Office of Population Census and Surveys for the planning and carrying out of population counts and surveys.

See also CENSUS TRACT.

ER MODEL

See ENTITY RELATIONSHIP MODEL.

ERROR

In GIS, errors may be associated with mistakes in either the spatial or ATTRIBUTE VALUES of the data. These may result from the original state of the data, from data input methods, or from functions and algorithms applied in the GIS.

ERROR PROPAGATION

The compounding of errors in computational results during data analysis. Such error propagation is brought about by errors in the data, by poorly designed algorithms, or by low numeric precision.

ES

See EXPERT SYSTEM.

EUCLIDEAN DISTANCE

The shortest distance between two points in a plane. Its value is found by resolving Pythagorus' theorum for a right angled triangle formed from the two points.

Euclidean distance (d) between two points which is calculated using Pythagorus' theorum.

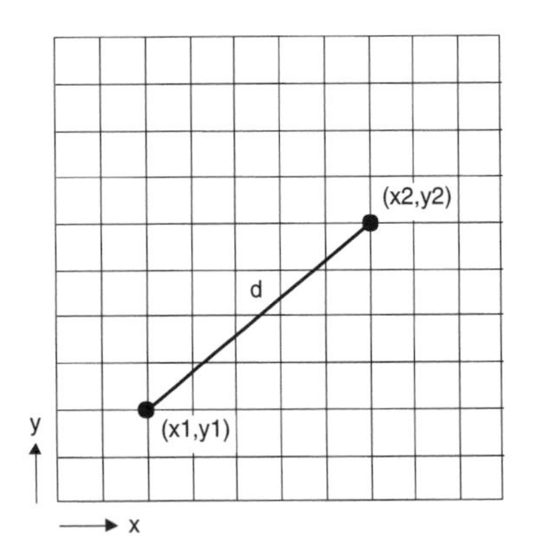

EUROGI

EUROPEAN UMBRELLA ORGANIZATION FOR GEOGRAPHIC INFORMATION An organization, under the auspices of the European Commission's DGXIII, which covers the GIS community in Europe and comprises national and pan-European sectorial associations with an interest in geographic information. It aims to stimulate, encourage, and support the development and use of geographic information at the European level.

EUROPEAN UMBRELLA ORGANIZATION FOR GEOGRAPHIC INFORMATION

See EUROGI.

EUROSTAT

The statistical agency of the European Communities.

EXPERT SYSTEM

ES A system designed for solving problems in a particular application area by drawing inferences from a knowledge base acquired by recording and structuring human expertise.

EXTRAPOLATION

A method or technique to extend data or inferences beyond known values.

FACE

A region bounded by a closed sequence of edges. Faces are contiguous and do not overlap.

FALSE COLOUR

A technique employed in the display of remotely sensed data to show an image of a waveband, such as near-infrared and thermal infrared images, which does not have a 'colour' visible by the human eye.

FEATURE

A group of spatial elements which together represent a real-world entity. Often used synonymously with the term object. A complex feature is made up of more than one group of spatial elements, for example, a set of line elements with the common theme of roads representing a road network.

FEATURE CODE

An ALPHANUMERIC code which identifies, describes, and/or classifies a set of spatial elements.

FEATURE EXTRACTION

The process of identifying features within a data set using pattern-recognition algorithms. *See also* AUTOMATED FEATURE RECOGNITION.

FEDERAL INFORMATION PROCESSING STANDARD

FIPS Data processing and transfer standards established by the US federal government to ensure that data created under federal funding can be shared by other government agencies.

FIELD

1) A set of ALPHANUMERIC characters comprising a unit of information.
2) A location in a data record in which a unit of information is stored. For example, in a database of addresses, one field would be 'city'.

FILE

A named set of computer records stored or processed as a unit.

FILE FORMAT

A structure for organizing data in a computer file. May be unique to a specific computer program or may conform to a general standard.

FILTER

Generally a 3 by 3 or 5 by 5 WINDOW of cells in which each cell is assigned a specific number. This number determines the WEIGHTING which the value of the corresponding cell in a raster will have during a filter operation.

An example of a 3 by 3 filter.

Total of cells is 1.

.05	.05	.05
.05	.6	.05
.05	.05	.05

FILTER OPERATION

An operation producing a new raster in which the value of each cell is based on its original one value and possibly that of adjacent cells combined in some mathematical manner. A WINDOW of cells is often used in such calculations.

FIPS

See FEDERAL INFORMATION PROCESSING STANDARD.

FLATBED PLOTTER

A device for creating HARD COPY graphic output. The paper or film is fixed to a flat surface which is traversed by the plotting device.

FLATBED SCANNER

A type of SCANNER in which the original is placed on a horizontal surface, and the scanner head moves over it in both x and y directions.

FLAT FILE

Data stored in a computer file which is arranged conceptually as a 2D array of data elements. Often used in reference to database programs in which each RECORD in the database contains the same set of FIELDS.

FLOATING POINT

A technique for representing numbers without using a set position for the decimal point. This improves the calculating capability of computer processors for arithmetic operations and enables the precision of results to be determined by the data processing.

FONT

A design type of ALPHANUMERIC characters and symbols. Some of the most commonly used are Helvetica, Times Roman, and Courier.

FORMAT

A systematic and repeatable arrangement of computer data.

See also TRANSFER FORMAT.

FOURIER ANALYSIS

A mathematical technique for transforming and breaking down data into a sets of sine and cosine waves. It is frequently used in oceanic and atmospheric applications. It is also used as a filtering technique in IMAGE PROCESSING to remove unwanted NOISE in the data or to reduce the size of the data set.

FRACTAL

An ENTITY which has variation within it that is similar at all scales, a characteristic known as self-similarity. Thus, at all scales, the entity exhibits the same amount of complexity. The best example is that of a coastline which shows new levels of complexity as you zoom in.

An example of a fractal object is the Von Koch curve.

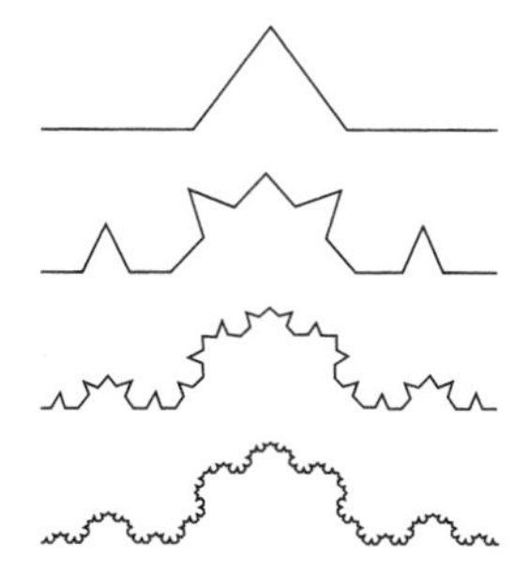

FUZZY ANALYSIS

A method for dealing with uncertainty and quantitative imprecision in data. Rather than depending on Boolean logic in which an ENTITY either is or is not a member of a union of two sets, a membership function is used which expresses the probability that a certain outcome will be realized.

GAP

The space between two elements on a digitized map which should intersect. It results from errors in the digitizing of the data.

GBF-DIME FILE

GEOGRAPHIC BASE FILE A geographic data-coding system developed to automate the processing of census forms in the US. It resulted in a series of TOPOLOGICALLY STRUCTURED DATA records for 350 major US cities which were current to 1980.
See also DIME and TIGER.

GCP

See GROUND CONTROL POINT.

GENERALIZATION

A simplification of data, so that information remains clear and uncluttered when the representation scale is reduced. It usually involves a reduction in detail, a resampling to larger spacing, or a reduction in the number of points in a line. In GIS it is an important consideration given the many scales at which data is available. It has implications for the digital representation, analysis, and display of the data.
See also AGGREGATION.

GEOCODE

A code associated with a spatial element which describes its location. For example, this might be a co-ordinate or a postcode.

GEOCODING

The process by which the geographic co-ordinates of a location are determined by its address, postcode, or other explicitly non-geographic descriptor.

GEODESY

The science of measuring the Earth.

GEODETIC DATUM

The definition of a particular spheroid and its position and orientation relative to the GEOID. Maps and GIS using different geodetic datums require complex transformations of the data being transferred between them.

GEOGRAPHIC BASE FILE

See GBF-DIME FILE.

GEOGRAPHIC DATA

Any information which includes a description of a location on or near the Earth's surface. This may include generic descriptions, for example, place names or particular geological strata.

GEOGRAPHIC INFORMATION SYSTEM

See GIS.

GEOGRAPHIC INFORMATION SYSTEMS ASSOCIATION

See GISA.

GEOID

The shape of the Earth as defined by mean sea level and its imagined continuation under the continents at the same level of gravitational potential.

GEOMATICS

A term coined in Canada to describe the field of activities which integrates all the means used to acquire and manage spatial data for the scientific, administrative, legal, and technical operations involved in the process of the production and management of spatial information.

GEOMETRY

In GIS, an ambiguous term often used to describe the manner in which a real-world ENTITY is represented geometrically in a database. Thus, the shape of an entity could be described in terms of its stored co-ordinates and the lines between these co-ordinates. For example, the geometry of a highway system might be described as a network or a set of lines and intersections or nodes.

GEOREFERENCE SYSTEM

A CO-ORDINATE SYSTEM for points on the Earth's surface. Examples of such a system are the Universal Transverse Mercator system (UTM) and the latitude and longitude graticule.

GEOSTATIONARY SATELLITE

A satellite which orbits the Earth at a distance of some 36 000 km above the Equator and which travels in the same direction as the rotation of the Earth. Its speed of movement keeps it positioned above the same hemispherical portion of the globe. Examples include the METEOSAT and GOES E and W.

GEOSTATISTICS

A set of statistical techniques which depend upon an analysis of spatial covariance, or how the value of an attribute varies through space. KRIGING, although only one aspect of geostatistics, is often used synonymously with it.

GIGABYTE

See BYTE.

GIS

GEOGRAPHIC INFORMATION SYSTEM

A computer system for capturing, managing, integrating, manipulating, analysing, and displaying data which is spatially referenced to the Earth.

GISA

GEOGRAPHIC INFORMATION SYSTEMS ASSOCIATION An academic society, based in Japan, to promote the development of GIS theory and application. Members include academic and other researchers, consultants, industry specialists, and students.

GKS

GRAPHICS KERNEL SYSTEM A set of functions for computer graphics programming, and for interfacing between an applications program and the graphical input and output devices.

GLOBAL POSITIONING SYSTEM

See GPS.

GOURAD SHADING

A technique used in hillshading which interpolates smooth transitions in the shading between the individual slope facets representing the elevation surface.

GPS

GLOBAL POSITIONING SYSTEM

A constellation of radio-emitting satellites deployed by the US Department of Defense used to determine location on the Earth's surface. Using a GPS receiver, it is possible accurately to determine one's location at or near the surface of the Earth without the need for any reference marks.

GRADIENT ANALYSIS

An analytical procedure using data in a DTM which allows the maximum rate of change between adjacent elements of a surface to be derived. It is generally used to determine the slope on a land surface.

GRAPHICAL USER INTERFACE

See GUI.

GRAPHIC PRIMITIVE

A basic graphic element that can be used to construct a display such as a point or line.

GRAPHICS KERNEL SYSTEM

See GKS.

GRAPHICS TABLET

A small digitizer used for interactive work such as required with some SPATIAL DATA handling software.

GRATICULE

An intersecting network of latitude and longitude lines drawn on a map or on the surface of the Earth. On a MAP PROJECTION the lines are not necessarily orthogonal or even, in general, straight.

The graticule for a conical map projection of north-west Europe.

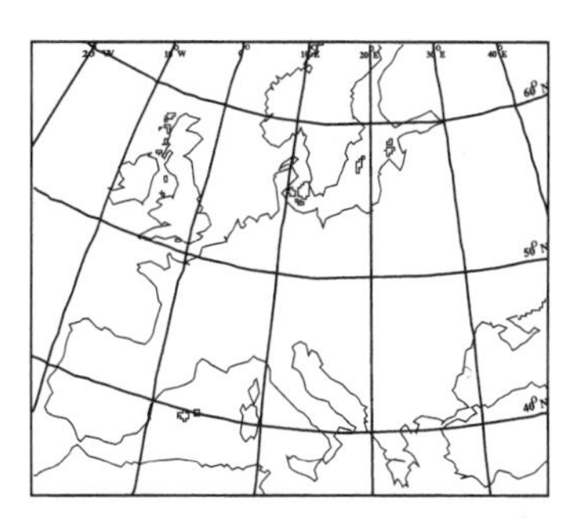

GREY SCALE

A range of intensities between black and white through shades of grey.

GRID

1) An orthogonal set of lines depicting a planar CO-ORDINATE SYSTEM.

2) A set of rectangular cells or of points arranged in columns and rows.

GRID CELL

A two-dimensional element that represents space in a regular or nearly regular TESSELLATION of an area. Usually represented by a square.

GRID REFERENCE

The position of a point on a map expressed in terms of co-ordinates of a planar CO-ORDINATE SYSTEM. Some countries, for example, France, New Zealand and the UK, have developed their own National Grid which is a system used to reference points within them.

GROUND CONTROL POINT

An easily identifiable feature with a known location which is used to give a geographic reference to a point within a map or remotely sensed image. It is often used in the geometric correction of spatial data sets.

GROUND TRUTH

See REFERENCE DATA.

GUI

GRAPHICAL USER INTERFACE An interface between the computer and the person using it, which makes use of icons, menus, and a pointing device to select options and to execute commands or graphs. It usually has the capability to display more than one WINDOW on the same screen.

HARD COPY

A print or plot of output data on paper or film media.

HARD DISK

A large capacity mechanical, magnetic computer storage device for programs and data. Generally installed internally in a computer and non-removable although external removable versions are available.

HARDWARE

The various physical components of an information processing system. The term refers to devices such as the computer, the screen, plotters, and printers.

HEADER

The initial lines in some data files which contain information on the data present (see METADATA). This might include the number of records and columns and the numerical format of the data.

HELMERT TRANSFORMATION

After DIGITIZING it is necessary to transform the SPATIAL DATA from DIGITIZER CO-ORDINATES to real-world ones prior to any analysis. The Helmert transformation converts digitizer to CARTESIAN

CO-ORDINATES through the following equations:

$$x_r = A + Cx_d + Dy_d$$

$$y_r = B + Dx_d + Cy_d$$

where x_r, y_r = Cartesian co-ordinates; x_d, y_d = digitizer co-ordinates and A, B, C, D are derived coefficients.

See also AFFINE TRANSFORMATION.

HEXTREE

A hierarchical raster data model based on hexagons. It is similar to the QUADTREE but the spatial units are hexagonal.

HIDDEN LINE REMOVAL

A technique which removes or masks a line or segment of a line in a 2.5D or 3D display of an area when it would ordinarily be obscured from view. This is of particular use in the creation of perspective views.

HIERARCHICAL DATA MODEL

A method of organizing SPATIAL DATA into elements based on a tree-type structure of different levels. It requires that space is divided into tessellating shapes that can be recursively decomposed into similar though smaller ones. Spatial elements may then be represented at a suitable level of detail. QUADTREE and HEXTREE are examples.

HIERARCHICAL DATA STRUCTURE

A method of organizing files and other data within a computer so that the units stored are referenced according to a tree-type structure.

HIGH-LEVEL LANGUAGE

A computer programming language, such as FORTRAN or C++, which uses English-like words or common mathematical symbols to define commands.

HIGHLIGHTING

A means of emphasizing a display element by modifying its visual attributes.

HILLSHADING

A technique for enhancing relief depiction in 2.5D or 3D displays of SPATIAL DATA. It attempts to emulate the way natural light low on the horizon illuminates the terrain. Also known as relief shading.

HISTOGRAM

A graphic representation of the frequency distribution of a set of data.

HYPERSPECTRAL SCANNERS

See IMAGING SPECTROMETRY.

ICON

A graphic symbol, displayed on a screen, that the user can point to with a device, such as a mouse, in order to select a particular function or software application.

IEEE

INSTITUTE OF ELECTRICAL AND ELECTRONICS ENGINEERING INC. A major international professional body and an accredited standards setting organization.

IGES

INTERNATIONAL GRAPHICS EXCHANGE SYSTEM An ANSI standard for the exchange in digital form of CAD drawings.

IGGI

INTER-DEPARTMENTAL GROUP ON GEOGRAPHIC INFORMATION A UK central government group which aims to provide a forum for government departments to develop a common view on issues related to geographic information, to facilitate the effective use of government-held geographic data, and to help overcome barriers to the use of such data.

IMAGE

A 2D digital ARRAY of values. Usually used in reference to remotely sensed scenes but also used for scanned maps, photographs, other rasters, and screen displays.

IMAGE CLASSIFICATION

A set of IMAGE PROCESSING techniques which applies quantitative methods to the values in a remotely sensed scene to group pixels with similar DIGITAL NUMBER values into feature classes.

IMAGE ENHANCEMENT

A set of IMAGE PROCESSING techniques used to accentuate the contrast between different features within a remotely sensed scene. This helps the visual interpretation of the elements within the image.

IMAGE PROCESSING

A set of digital techniques used to create, analyse, enhance, interpret, or display image data.

IMAGE RECTIFICATION AND RESTORATION

A set of IMAGE PROCESSING techniques for reducing or eliminating distortions and noise in the data resulting from the sensing, collecting, and data transferring systems and processes.

IMAGING SPECTROMETRY

Remote sensing using instruments which have many, often hundreds of detectors, that are sensitive to very narrow wavebands within the ELECTROMAGNETIC SPECTRUM. They sense principally within the visible, near-infrared, and mid-infrared parts of the spectrum. The resulting images are particularly important for geological studies.

IMPORT

The process of bringing data or programs from one computer system into another.

INCLINED CONTOURS

A method used to give a better 3D impression of relief. It represents intersections of the surface with parallel inclined planes that are at right-angles to the direction of the assumed illumination.

INFORMATION

Intelligence resulting from the assembly, analysis, or summary of data into a meaningful form.

INHERITANCE

A term that describes the transfer of characteristics from one level of data or representation to another, lower linked level. It is a concept used extensively in OBJECT-ORIENTED programming and representations of space.

INK-JET PLOTTER

A colour printing device in which electrostatically charged ink drops are finely sprayed onto the paper or film.

INPUT

The process of entering data into a computer. The term is also used to describe the actual data that are to be entered. One form of input is digitizing.

INSTITUTE OF ELECTRICAL AND ELECTRONICS ENGINEERING INC.

See IEEE.

INTEGRATION

The combining of data of different types and from different sources and systems. It may also be used to refer to the integration of computer systems so that data transfer between them can be performed without any operator intervention.

INTERACTIVE

Describes the two-way communication between the computer and the user in which responses from one are immediately relayed to the other, either through the keyboard or by display on the screen.

INTERACTIVE DIGITIZING

A mode of DIGITIZING in which two-way communication is possible between the operator and the computer. As the data are entered by the operator, it is immediately displayed on the computer screen, so allowing the operator to see what has already been captured. Most digitizing software supports this mode of operation.

INTERACTIVE EDITING

A form of editing interface in which maps and data are displayed while data are edited. The operator can see immediately the effect of any edit in the database.

INTER-DEPARTMENTAL GROUP ON GEOGRAPHIC INFORMATION

See IGGI.

INTERFACE

The junction which allows the linking together of two or more computer components. This, for example, might be between software and hardware, between hardware and hardware, or between human operator and software.

INTERIOR AREA

An AREA not including its BOUNDARY, for example, the inside of a POLYGON.

INTERNATIONAL GRAPHICS EXCHANGE SYSTEM

See IGES.

INTERNATIONAL STANDARDS ORGANIZATION

See ISO.

INTERNET

An international network of dispersed local and regional computer networks used predominantly for sharing information and resources. Developed primarily for military and then academic use, it is now accessible through commercial on-line services to the general public.

INTERPOLATION

A series of techniques and algorithms used to estimate ATTRIBUTE VALUES for areas that are unsampled, based on known DATA at surrounding sample sites. Examples include techniques such as KRIGING and THIESSEN POLYGONS.

INTERSECTION

The point at which one line crosses another. In GIS there are a number of contexts in which this is used, for example, in polygon overlay analysis to create the intersection of two area features.

Intersection of two lines.

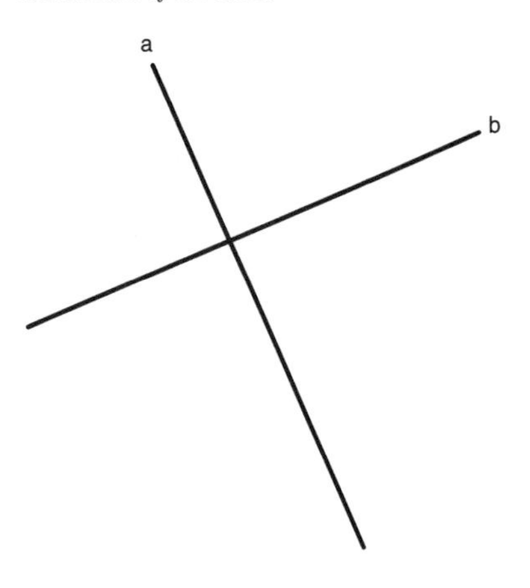

INVISIBLE LINE

A line which is sometimes used to make a logical connection between two non-contiguous parts of a feature or between two different features but which is not displayed.

IRRADIANCE

The total amount of incoming ELECTROMAGNETIC ENERGY radiated onto a surface. It is usually measured in watts per square metre.

IRREGULAR TESSELLATION

See TESSELLATION.

ISLAND

A closed 2D spatial element which lies completely within another.

An island polygon of a lake within a forest polygon.

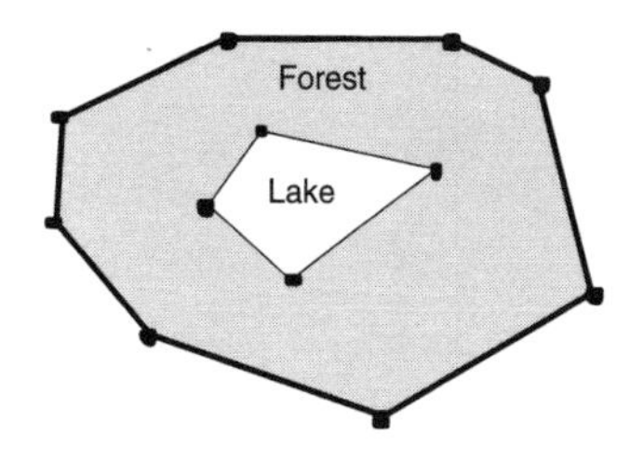

ISO

INTERNATIONAL STANDARDS ORGANIZATION A worldwide federation of national standards bodies that defines rules, critera, or measurements that are to be adopted as international standards.

ISOCHRONE

A type of ISOLINE in which the lines join points of equal travel time from a particular point. Commonly used in route-planning applications.

ISOLINE

A line joining points of equal value.

ISOLINE MAP

A map with the surface features shown as a series of lines joining points of equal value. The common example is the contour map where lines join points of equal elevation. It is also known as an isopleth map.

ISOPLETH MAP

See ISOLINE MAP.

JOIN

The process of connecting two or more separately digitized spatial ENTITIES. If two line segments are joined, they become one spatial object for any further processing.

JUSTIFICATION

In GIS, the position of a text string or symbol on a map relative to the location at which it has been spatially referenced.

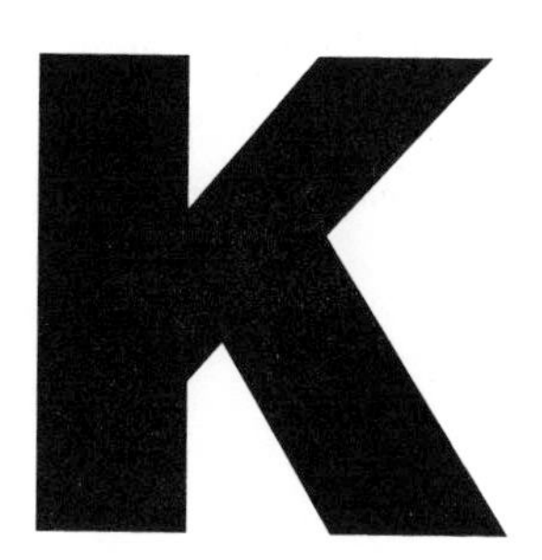

KB

See BYTE.

KEY

A unique identifier for each RECORD in a table or each ENTITY in a data set which is used to link this table or data set to others.
See also RELATIONAL JOIN.

KRIGING

An INTERPOLATION technique based on numerical measurements of the spatial variation of known points different distances apart. It is used predominantly in GIS work to obtain estimates of surface elevation from a set of known points.
See also GEOSTATISTICS.

LABEL

An ALPHANUMERIC element, used with the VECTOR DATA MODEL, that displays attribute values or identifiers for points, lines, or polygons.

LAMBERTIAN REFLECTION

See DIFFUSE REFLECTION.

LAND COVER

The surface materials, such as crops and water, that are found in an area.

LAND INFORMATION SYSTEM

See LIS.

LAND PARCEL

A unit of land, usually delineated according to land ownership or land use.

LANDSAT

A series of American POLAR-ORBITING SATELLITES which collect data suitable for earth surface mapping and resource management.

LAND TERRIER

A document system comprising a set of annotated maps and ledgers containing textual information to record land and property.

LASER PRINTER

An output device which uses laser technology, similar to that used in photocopiers, to print text and images onto paper.

LATITUDE

The angular distance north or south between a point on the Earth's surface and the Equator. The distance is measured with reference to an idealized, spheroid-shape of the Earth.

LAYER

A usable subdivision of a data set, generally containing elements of a particular theme. The subdivisions are registered to a common CO-ORDINATE SYSTEM, thus enabling analysis and integration across the various themes.

The layer concept.

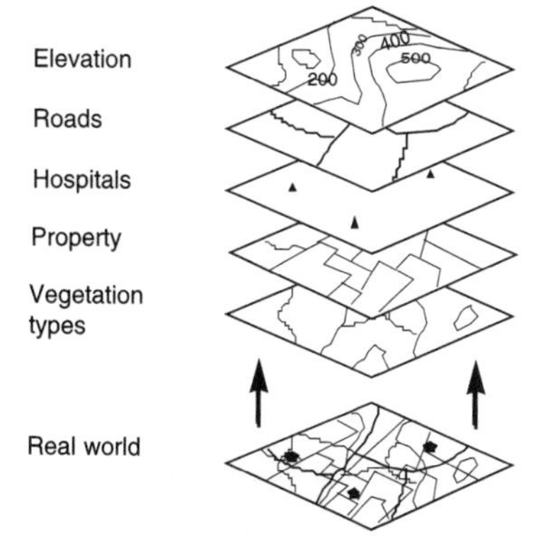

LEAST-COST SOLUTION

See OPTIMAL PATH SELECTION.

LEGEND

A part of a map which contains explanations of the symbols, colours, and shading that have been used to code the various elements and data values.

LICENCE

In common GIS usage, licence refers to the legal agreement between a software company and the user which determines how that software may be used. Licences generally limit the number of computers on which a single software package can be used.

LINE

A series of connected, co-ordinate points forming a simple linear feature. It is a fundamental spatial unit in the VECTOR DATA MODEL. It is used to represent features, such as rivers and roads, or to form the boundaries of polygons. It is identical to an arc.

LINEAGE

The ancestry of a data set describing its origin and the processes by which it was derived from that origin.

LINE FOLLOWING

An algorithm which is capable of tracing lines from an image, scanned map, or other raster data, converting them into vector form.

LINE-IN-POLYGON OPERATION

A GIS operation used to determine which lines of a data set are contained in a particular polygon of another data set. For example, it might determine which roads are contained in a particular administrative area.

A line-in-polygon operation would detect whether lines of highways a or b pass through the forest polygon.

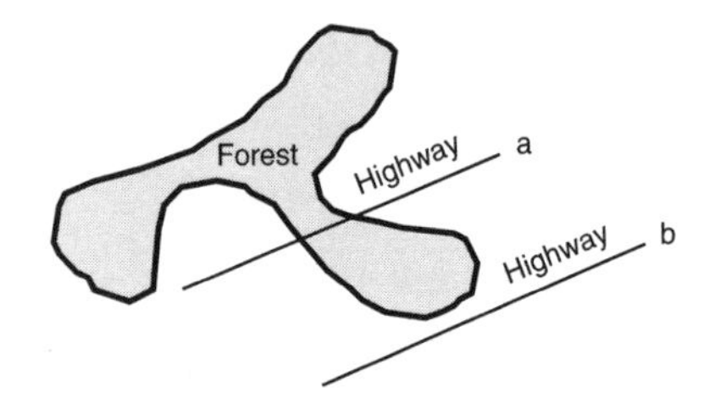

LINE SMOOTHING

An automatic process which removes irregularities and roughness from a digitized line within a TOLERANCE distance established by the operators. The process causes some or all of the co-ordinate points of the line to be changed.

LINE THINNING

1) A process in which the number of co-ordinate points defining a line is reduced while still preserving the general shape of it. The term weeding is also used to describe this process.

Line thinning for a reduction in the number of points used to define a line.

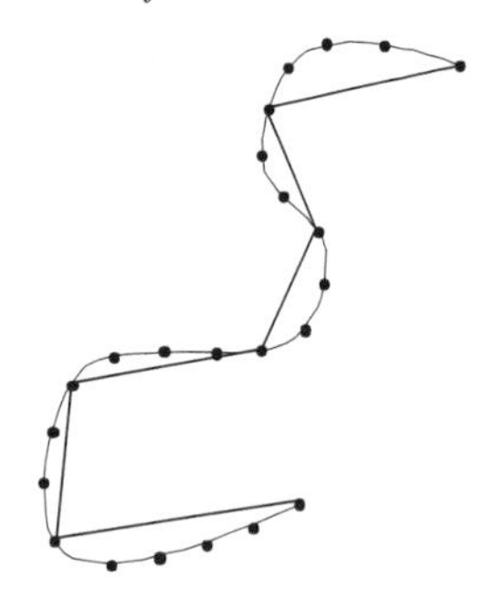

2) A process in which the number of pixels used to represent the width of a linear feature in a raster data set is reduced to the minimum number required to produce a contiguous series of cells.

Line thinning for a reduction in the number of pixels used to define a line.

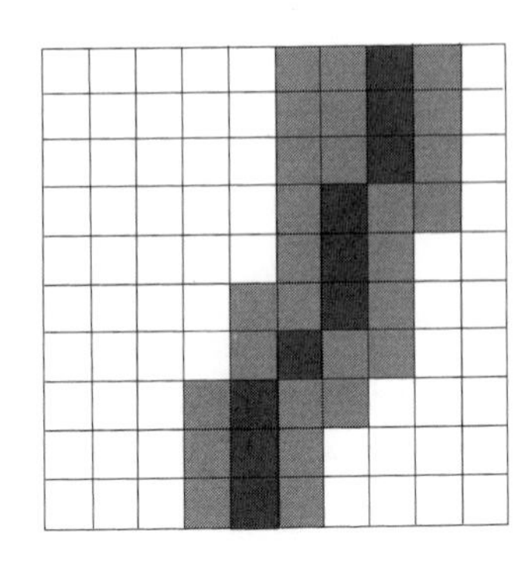

LINK

A line without logical intermediate intersections. In computer programming, it refers to the connecting of individual programs or modules into a larger executable set.

LIS

LAND INFORMATION SYSTEM A system for capturing, storing, checking, integrating, manipulating, analysing, and displaying data about land and its use, ownership, and development.

LOCATION-ALLOCATION

A procedure which is used to assign a set of facilities to a set of regions or locations according to some optimal criteria. For example, location-allocation could be used to determine the optimal locations for ambulance garages in order that travel time to any potential accident location is minimized.

LOCATIONAL REFERENCE

A method for relating data to a specific spatial position.

LOGICAL EXPRESSION

A mathematical expression which combines logical and BOOLEAN OPERATORS and results in a value of true or false.

LOGICAL OPERATOR

One of the set of mathematical operators including less than (<), greater than (>), equal to (=), not equal to (<>) and various combinations of these. Logical operators are often combined with BOOLEAN OPERATORS in a logical expression.

LONGITUDE

The angular distance of a point east or west of an arbitrarily defined meridian, usually taken to be the Greenwich meridian. The distance is measured with reference to an idealized, spheroid shape of the Earth.

LOOKUP TABLE

A table which contains a key field and related values which can be linked to other data tables to provide additional ATTRIBUTE VALUES about the ENTITIES.

LOOSELY COUPLED MODELS

A term which refers to the linking of a GIS to a mathematical model through common data elements alone. Any processing and analysis is undertaken separately within each system.

LOW-LEVEL LANGUAGE

A programming language in machine code, which allows direct communication with the computer processor. Often uses binary or hexadecimal code and requires a detailed knowledge of the structure of computer memory and the fundamental processing commands. Few computer users today use low-level, or even HIGH-LEVEL LANGUAGES, choosing instead to control computers through applications programs with graphical user interfaces (GUI).

MACRO

A series of instructions which together perform a particular task, and which may be executed by a single command. *See also* CUSTOMIZATION.

MAINTENANCE

In common GIS usage, maintenance refers to the full suite of operations which is required to keep a database and/or a commercial software product current. Software maintenance generally involves a contract between the software vendor and the user for which the user pays an annual fee.

MANHATTAN DISTANCE

This is a metric system based on a grid. The distance between two points is defined in terms of the rectangular distance or the number of grid cells in each direction.

Manhattan distance (d) between two points.

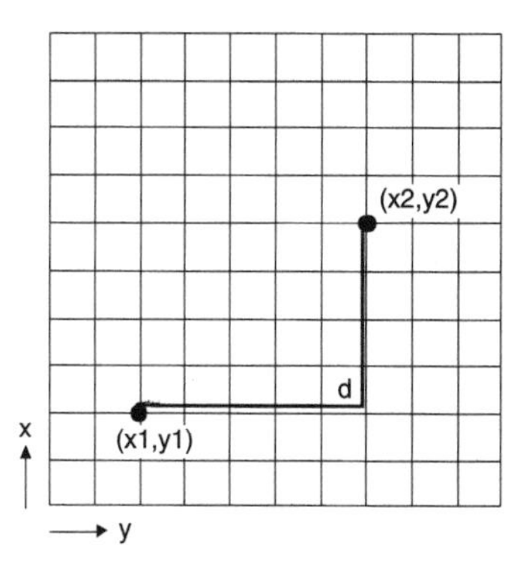

MAP

A graphic representation of geographically distributed phenomena. The information displayed may be in the form of symbols or signs. Accuracy and detail are functions of map projection and scale.

MAP GENERALIZATION

See GENERALIZATION.

MAP PROJECTION

A method by which the curved surface of the Earth is portrayed on a flat surface. This generally requires a systematic mathematical transformation of the Earth's graticule of lines of longtitude and latitude onto a plane.

Azimuthal, conical, and cylindrical based map projections.

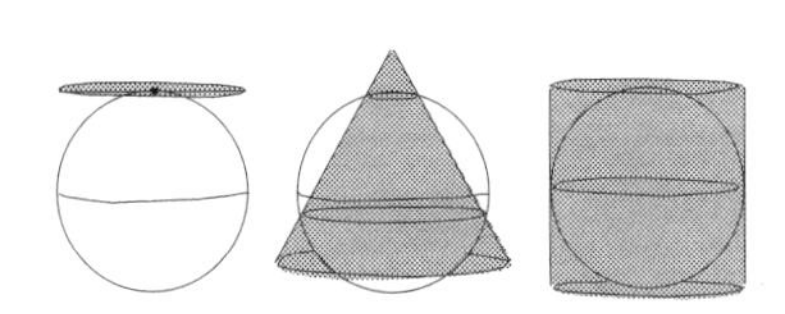

MAP SCALE

See SCALE.

MAXIMUM LIKELIHOOD CLASSIFIER

An IMAGE CLASSIFICATION technique. Given a known data set which has already been classified, for each class the variance, covariance, and mean values of each waveband are calculated. The resulting probability functions are then used to allocate other image pixels to the classes to which they have the maximum likelihood of belonging.

Maximum likelihood classification technique with contours showing the probability of likelihood that an unclassified pixel will belong to the different land cover types; inner contours of the different groups have the greatest probability values.

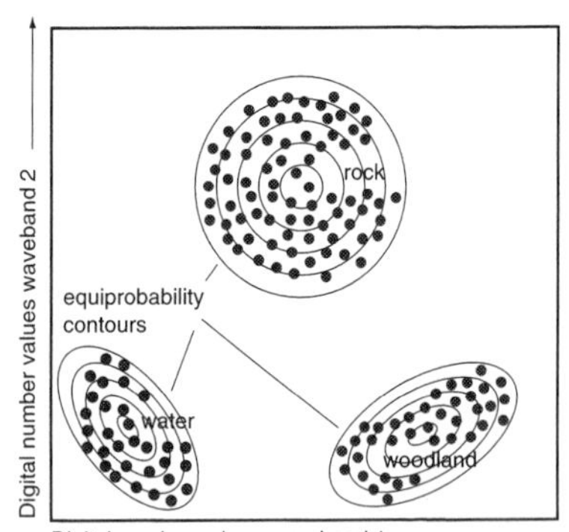

MEGABYTE

See BYTE.

MEGRIN

MULTI-PURPOSE EUROPEAN GROUND RELATED INFORMATION NETWORK

A consortium of European national mapping agencies, under the auspices of CERCO, to allow the geographic data produced by those mapping agencies to be more widely available. Projects include the development of joint topographical data such as national and administrative boundaries.

MENU

A list of options displayed by a data processing system, from which the user can select an action to be initiated. These choices may be displayed in the form of ALPHANUMERIC text or ICONS.

MENU BAR

An area along one edge of a window within a computer package used to display names or icons for a menu.

MERCATOR PROJECTION

A cylindrical MAP PROJECTION centred along the Equator in which lines of latitude are represented as straight lines and intersect the lines of longitude at right angles. The main distortions are to the areal properties of spatial elements in the high latitudes. A transverse Mercator projection is similar although centred along a line of longtitude.

The Earth shown using the Mercator projection.

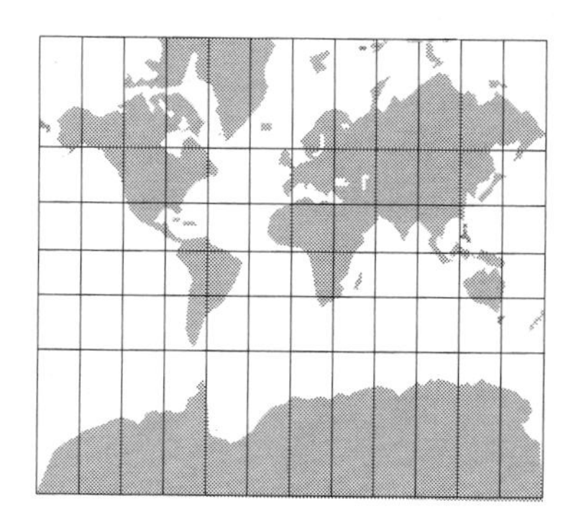

METADATA

Information about data. Examples are data quality information, currency, LINEAGE, ownership, and feature classification information.

METEOSAT

A European meteorological GEOSTATIONARY SATELLITE. Its position gives good coverage of the whole of Africa as well as Europe.

MINIMUM DISTANCE TO MEAN CLASSIFIER

An IMAGE CLASSIFICATION technique which first calculates the mean value for each of the classes within each waveband in a known data set. The values of each pixel in the image are compared to these mean values and then allocated to the class whose mean is the closest (minimum distance).

Minimum distance to mean classification technique with distances shown from an unclassified pixel to the mean values of waveband reflectance for rock, crop, water, and woodland land cover types.

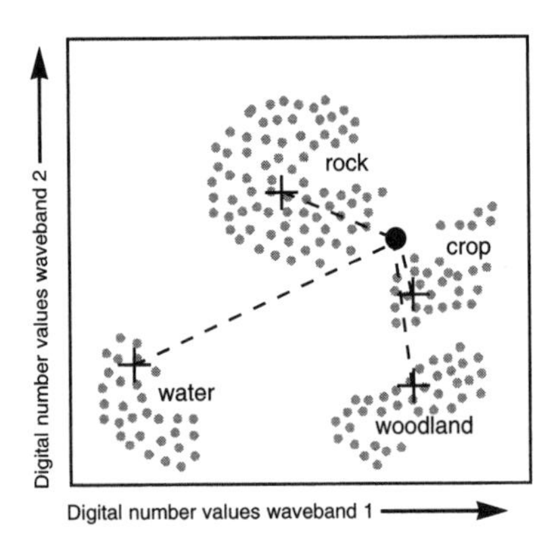

MINIMUM MAPPING UNIT

MMU The dimension of the smallest unit represented in a data set. For example, on a soils polygon map, the minimum mapping unit might be two hectares.

MMU

See MINIMUM MAPPING UNIT.

MODEL

An ABSTRACTION and description of the real world or part of it.

MORTON ORDER

An identification system which assigns a unique code to each node in a QUADTREE so that its geographic location can be determined. It is a system adopted in a number of quadtree-based GIS. Also called Peano keys.

MOSAIC

A large image assembled from several contiguous raster data sets.

MOUSE

A hand-operated device used for pointing at and selecting areas on a computer screen. The location of the pointer on the screen is manipulated by moving the mouse over another surface.

MSS

MULTI-SPECTRAL SCANNER

A scanning device on board the early Landsat series of satellites. It collects data in the visible and infrared parts of the ELECTROMAGNETIC SPECTRUM.

MULTI-MEDIA

A combination of several communication media such as still and moving pictures, sound, graphics, and text.

MULTI-PURPOSE EUROPEAN GROUND RELATED INFORMATION NETWORK

See MEGRIN.

MULTI-SPECTRAL

The collection of remotely sensed data in two or more wavebands.

MULTI-SPECTRAL SCANNER

See MSS.

NAME PLACEMENT

See AUTOMATED NAME PLACEMENT.

NATIONAL CENTER FOR GEOGRAPHIC INFORMATION AND ANALYSIS

See NCGIA.

NATIONAL GRID

See GRID REFERENCE.

NATIONAL TRANSFER FORMAT

See NTF.

NCGIA

NATIONAL CENTER FOR GEOGRAPHIC INFORMATION AND ANALYSIS A US-based research centre predominantly dedicated to advancing the theory, methods, and techniques of geographic information analysis using GIS. Funded by the US National Science Foundation, it has sites at the University of California Santa Barbara, University of Maine, and the State University of New York at Buffalo.

NDVI

NORMALIZED DIFFERENCE VEGETATION INDEX An IMAGE PROCESSING technique which helps to highlight the presence or absence of vegetation. It is calculated by dividing the difference in values between NIR and red bands, by the sum of the two bands.

NEAREST NEIGHBOUR SAMPLING

An INTERPOLATION technique used with raster data which derives values for a new raster cell based on differences between the value of the original cell and the values held in the four nearest cells of the old raster data set.

NETWORK

1) A geometric or logical arrangement of nodes and interconnecting lines.
2) A database structure in which the links and relations between various data are explicitly defined.
3) A group of linked computers which are able to share software, data, and various hardware devices such as printers.

NODE

An intersection point where two or more arcs or lines meet.

NOISE

Random variations or error in a data set.

NORMALIZED DIFFERENCE VEGETATION INDEX

See NDVI.

NTF

NATIONAL TRANSFER FORMAT British Standard (BS 7567) for the transfer of geographic data, administered by the AGI. *See also* SDTS.

OBJECT

In recent years this has taken on a more specific usage with reference to GIS. It is a digital representation of a discrete spatial entity. An object may belong to an object class and will thus have ATTRIBUTE VALUES in common with other defined elements.

OBJECT CLASS

A set of objects with a specific theme such as 'stream' or 'woodland'. A hierarchical approach may be adopted in defining these sets, and ATTRIBUTE VALUES may be inherited from levels above a particular class.

OBJECT ORIENTED

A method of organizing data, commands, operations, and real-world elements whereby they are encapsulated into various discrete objects. This term has been applied to databases, programming languages, and recently to GIS.

OCTREE

A data structure used to minimize storage for 3D raster data. It is similar to the QUADTREE used in 2D spatial data.

OPTIMAL PATH SELECTION

An analytical procedure using network data in which the best route from one point to another is determined in terms of time, distance, cost, etc. This is often used on road data sets. Also known as the shortest-path or least-cost solution.

ORIGIN

The reference point (0,0) from which co-ordinates are measured.

OUTPUT

The result of GIS processes and analysis. It might be in a number of forms such as a computer screen display, a printed map, or tables of values.

OVERLAP

An area that is spatially common to two data sets or polygons.

OVERLAY

The superimposition of two or more data sets that are registered to a common CO-ORDINATE SYSTEM. If the data is in the raster form, then the result will be one raster data set whose cell values are some mathematical or logical combination of the values in the original data sets. If the data is in the vector form, then the result will be a new set of polygons which are formed by the intersection of all the boundaries in the original data sets. Most GIS provide such a function.

OVERSHOOT

The projection of a line beyond the true point of intersection with another. Also known as a dangle.

An overshoot.

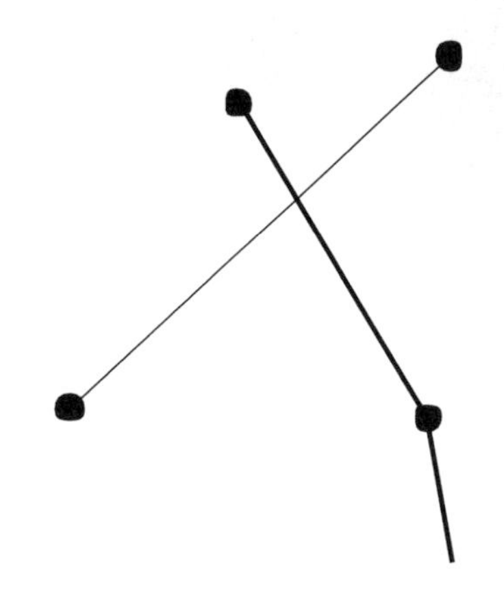

PACKAGE

A set of computer programs that can be used for a particular generalized class of APPLICATION.

PALETTE

The set of colours chosen to represent specific colour codes or digital numbers in a graphic display.

PARALLELEPIPED CLASSIFIER

An IMAGE CLASSIFICATION technique which uses a parallelepiped shape to define the range of DIGITAL NUMBER values for feature classes within a known data set. The parallelepiped represents a multi-dimensional space defined in terms of values for different wavebands. Pixels from an image are then classified according to which parallelepiped they fall into. Pixels which fall outside are classified as unknown.

Parallelipiped classification technique with shapes defined for different land cover types.

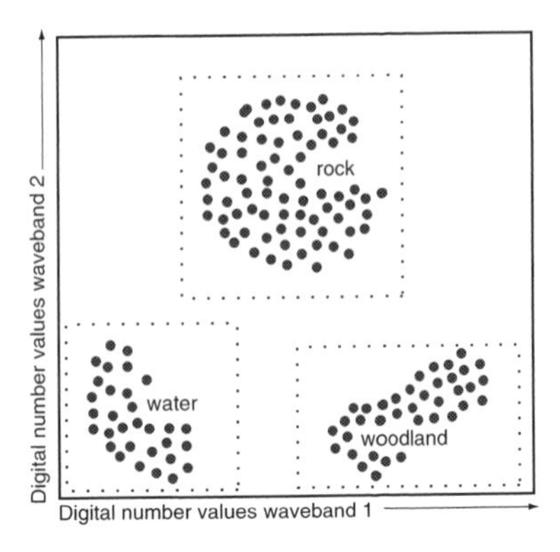

PARCEL

See LAND PARCEL.

PASSIVE SENSOR

A sensor which measures naturally available ELECTROMAGNETIC ENERGY. Most satellite sensors fall into this category.

PATTERN DETECTION

A technique used to determine the occurrence of a particular combination of shapes, patterns, colours, or features.

PECKED LINE

A line drawn as a series of dashes.

PEANO KEYS

See MORTON ORDER.

PEN PLOTTER

A drafting device for line drawings which mechanically moves an ink pen over the drawing surface.
See also FLATBED PLOTTER.

PERIPHERAL

Any device which is attached to the main computer system unit. PLOTTERS and DIGITIZERS are examples.

PERSPECTIVE VIEW

An operation in a GIS which enables the operator to generate a 2.5D or 3D representation of an area as if viewed from an angle other than the vertical.

PHIGS

PROGRAMMER HIERARCHICAL INTERACTIVE GRAPHICS SYSTEM

A standard (ISO 9592) set of graphics support functions to control the definition, modification, storage, and display of hierarchical graphics data.

PHOTOGRAMMETRY

The modelling of aerial photographic images for the purpose of extracting accurate information, particularly for the creation of maps.

PIECEWISE

Used to refer to a set of digital representations or mathematical functions which individually describe a geographically contiguous portion of the data set and which collectively describe the entire data set.

PIT

A depression in the digital surface of a DTM. Where this is incorrect, it often results from NOISE in the input data set, or from the INTERPOLATION of the original data. Pits are very disruptive in any processing which seeks to extract flow patterns. Also known as a sink.

PIXEL

1) The smallest element of a display device, such as a video monitor, that can be independently assigned attribute, such as colour and intensity.
2) In remote sensing, refers to the fundamental unit of data collection. A pixel is represented in a remotely sensed image as a rectangular cell in an array of data values. The term is an abbreviation for 'picture element'.

A pixel within a grid cell arrangement.

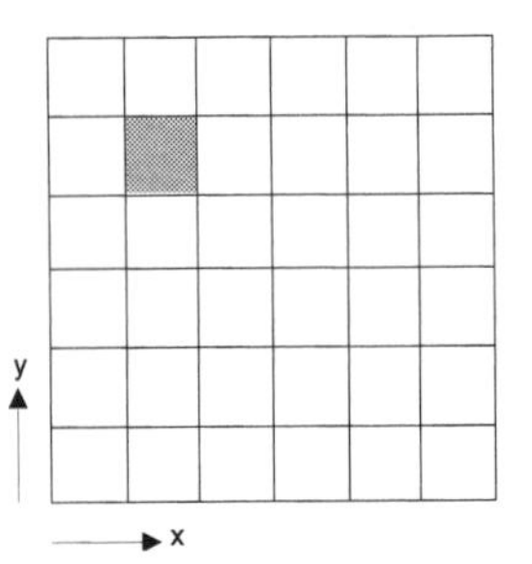

PLANAR ENFORCEMENT

A rule, applied to VECTOR DATA MODELS, that spatial elements may not overlap and must utilize all the space so that every place lies inside exactly one area.

PLOTTER

A drawing device for maps and graphic figures producing HARD COPY.
See also FLATBED PLOTTER and ELECTROSTATIC PLOTTER.

POINT

A zero-dimensional ABSTRACTION of an object, with its location specified by a set of co-ordinates.

POINT DIGITIZING

A method of digitizing where point locations are only recorded when a button is pressed on the PUCK.

POINTER

A symbol displayed on a screen, which a user can move with a pointing device, such as a MOUSE, to select items.

POINT-IN-POLYGON OPERATION

A GIS operation used to determine which points of a data set are contained in a specific polygon of another data set.

A point-in-polygon operation would detect whether points a or b fall in the polygon.

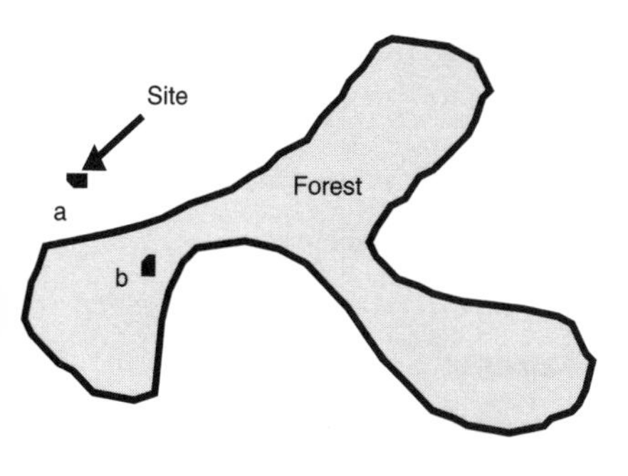

POLAR CO-ORDINATES

A 2D spatial referencing system which defines a point location in terms of the distance and horizontal angle from the origin.

A polar co-ordinate defined in terms of its angle and distance from a defined origin.

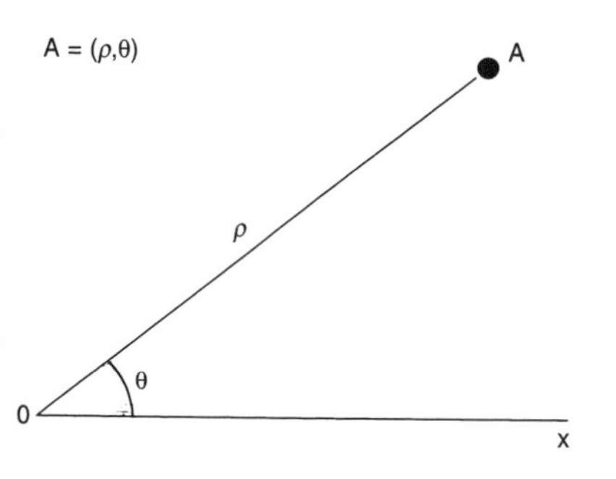

POLAR-ORBITING SATELLITE

A satellite whose orbital path around the Earth crosses over or near to the North and South Poles. It travels relatively close to the globe at an altitude of between 600–15 000 km. The term sun-synchronous is also used to describe such a satellite as it always crosses the Equator at exactly the same sun-time. Examples include the LANDSAT and SPOT series of satellites.

POLYGON

An area bounded by a closed line. It is used to describe spatial elements, such as housing and industrial units, administrative and political districts, and areas of homogeneous land use and soil types.

POLYGONIZATION

A process which creates polygons from line data. It ensures that lines meet exactly to enclose an area and also establishes the TOPOLOGY of the data set. It is a stage in PLANAR ENFORCEMENT.

POLYGON OVERLAY

See OVERLAY.

POLYLINE

A CAD term for a line.

POP-UP WINDOW

A WINDOW that appears on the display surface in response to some action.

PORTABILITY

The capability of a program to be executed on various types of computer systems without converting the program to a different language and with little or no modification.

POSITIONAL ACCURACY

A measure of the numerical difference between the measurement or representation of the position of a point and its true position.

POSTCODE

A coding system for referencing all properties in a country which have a postal address. The term is used generally, except in the US where the more specific term zipcode is used. The UK, for example, is divided into 120 areas, each area is divided into districts, each district into sectors, each sector into units. A unit postcode applies to a group of addresses of (approximately 15) neighbouring properties. It does not define an area.

PRECISION

The exactness with which a value is expressed, whether the value be right or wrong. It usually refers to the number of decimal places used to store the value in the database.

PRE-DEVELOPMENT MAPPING

A concept in which one party prepares a definitive digital map of a proposed development. The map is then used by all interested parties for detailed planning and recording purposes until the development has been surveyed on completion.

PREMCODE

A set of ALPHANUMERIC characters which, when added to the unit postcode, serves to identify a specific premise or address in the UK.

PROGRAMMER HIERARCHICAL INTERACTIVE GRAPHICS SYSTEM

See PHIGS.

PROJECTION

See MAP PROJECTION.

PROTOCOL

A set of communication standards which determines the language and codes that components of a system use to communicate with each other.

PROXIMITY

A type of analysis undertaken in a GIS to determine which spatial elements are the closest to an identified point, line, polygon, or grid cell.

PUCK

A pointing device that is positioned manually on the pad of a GRAPHICS TABLET or DIGITIZER to register the location of input points when digitizing locations.

PULL-DOWN MENU

A list of functions or commands that appears below the menu bar when the user selects a name or ICON from it.

PUSHBROOM SCANNING

See ALONG-TRACK SCANNING.

QUADTREE

A DATA STRUCTURE used to reduce the storage requirements of a raster by coding contiguous homogeneous areas as a single unit. The raster is recursively subdivided into four equal areas. Subdivision continues until all quadrants are homogeneous with respect to a selected attribute, or until a predetermined cut-off depth is reached. The data is then coded into the database with reference to the level of division a quadrant has reached.

The quadtree representation of space.

Area as represented on a map

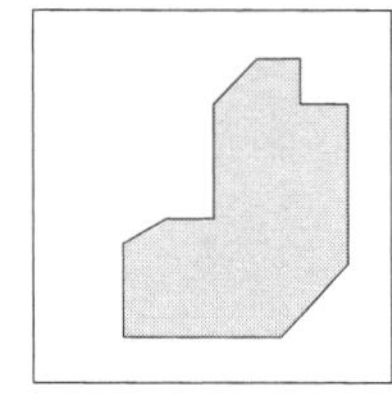

Quadtree representation

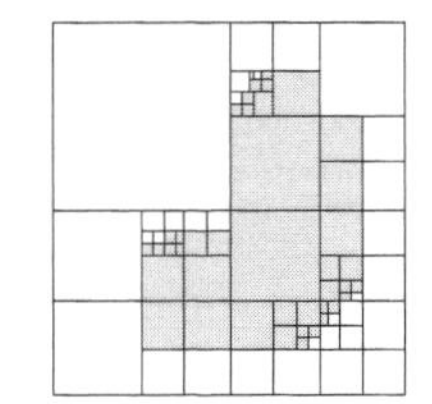

QUERY LANGUAGE

A computer language used in database management systems to retrieve, modify, add, or delete data. A common example is SQL (Structured Query Language).

RADIOMETER

An instrument for measuring ELECTROMAGNETIC ENERGY radiation within a waveband.

RADIOMETRIC RESOLUTION

The smallest difference in the amount of ELECTROMAGNETIC ENERGY radiation detectable by a sensor. A high radiometric resolution means that subtle changes in radiation levels may be discriminated.

RASTER DATA MODEL

The representation of SPATIAL DATA as a matrix of cells holding values for an attribute. The spatial position of an element is implicit in the ordering of the grid cells.

The raster data model.

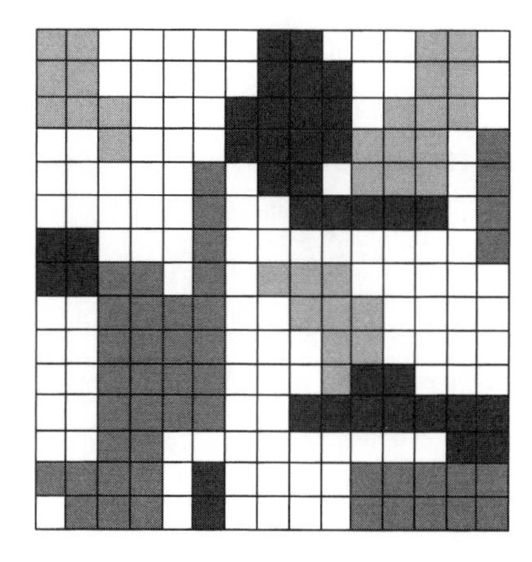

RASTER SCAN

A technique for generating or recording the elements of a display image by means of a line-by-line sweep across the entire display space. A common example is the generation of a picture on a television screen.

RASTER-TO-VECTOR CONVERSION

A process to convert an image made up of grid cells into a data set of lines and polygons.

RDBMS

RELATIONAL DATABASE MANAGEMENT SYSTEM A database management system that organizes data into a series of records held in linked tables. It allows different relations between the various records, fields, and tables to be established and is used to aid in data access and transformation.

REAL-TIME SYSTEM

A computer system that is able to receive continuously changing data from external sources and to process that data sufficiently rapidly to be capable of displaying data or images of current conditions.

RECORD

A set of related data fields describing a single ENTITY which is grouped as a unit for processing. For example, in an address database, the fields which together provide the address for a specific individual comprise a record.

RECTIFICATION

The removal of geometric distortion from an image or map. It is often used to bring several distorted images into a common CO-ORDINATE SYSTEM.

REFERENCE DATA

The field collection of data which is used in the interpretation of information gathered from other sources such as remote sensing. Also known as ground truth.

REFLECTION

In remote sensing the term refers to the return of waves of ELECTROMAGNETIC ENERGY by atmospheric or surface matter. Variations in reflection patterns help to distinguish between different materials. *See also* DIFFUSE REFLECTION and SPECULAR REFLECTION.

REGIONAL RESEARCH LABORATORY

See RRL.

REGISTRATION

The process of geometrically aligning images or maps so that corresponding features or cells are coincident.

REGULAR TESSELLATION

See TESSELLATION.

RELATIONAL DATABASE MANAGEMENT SYSTEM

See RDBMS.

RELATIONAL JOIN

An operation by which two data tables are related through a common field (KEY).

RELATIVE CO-ORDINATE

A co-ordinate identifying the position of an addressable point with respect to another addressable point without reference to its actual spatial location.

See also ABSOLUTE CO-ORDINATE.

RELATIVE POSITIONAL ACCURACY

The measure of the internal consistency of the positional measurements in a data set. For many local area purposes, for example, records of utility infrastructure, relative accuracy is more important than absolute accuracy. In this case, accurate measurement of offsets from fixed points is required rather than knowledge of the true geographic position.

RELIEF SHADING

See HILLSHADING.

REMOTE SENSING

The technique of obtaining data about the environment and the surface of the Earth from a distance, for example, from aircraft or satellites.

RENDERING

The conversion of the geometry, colouring, texturing, lighting, and other characteristics of ENTITIES stored in a data set into a display image in such a way that it has the characteristics of a photograph of a 3D object.

REPEATABILITY

The ability of a device to perform the same action consistently or to provide the same data given identical conditions. Given identical inputs, it may be possible to identify the limits within which the output will fall with a given statistical confidence.

RESAMPLING

The process of interpolating cell values from one raster data set into a new one which has larger or smaller cells or which may not be registered with it.

RESOLUTION

A measure of the ability to detect variation. High resolution implies a high degree of discrimination but has no implication as to accuracy.

See also SPATIAL RESOUTION.

RLE

See RUN-LENGTH ENCODING.

RRL

REGIONAL RESEARCH LABORATORY

A series of research centres established in the UK dedicated to GIS research and the demonstration of practical applications.

RUBBER BANDING

See RUBBER SHEETING.

RUBBER SHEETING

A procedure in which the co-ordinates of all the data points in a data set are adjusted to better match the known locations of a few points in the set. Connections between elements in a data set (TOPOLOGY) are retained although relative distances between points may be altered through the stretching, shrinking, or reorienting manipulations needed to meet the new geometric constraints. The procedure is often used in the registration of one map or image to another and is a form of CO-ORDINATE TRANSFORMATION. It is also known as rubber banding.

RUN-LENGTH ENCODING

RLE The process of encoding a sequence of numeric values in terms of the number of successive digital data elements that have the same value. This method is applied to raster data to reduce its storage requirements.

Run-length encoding records the start and end of the occurrence of a particular class for each row, so for row e the values would be 5, 6 and for row f 4, 7.

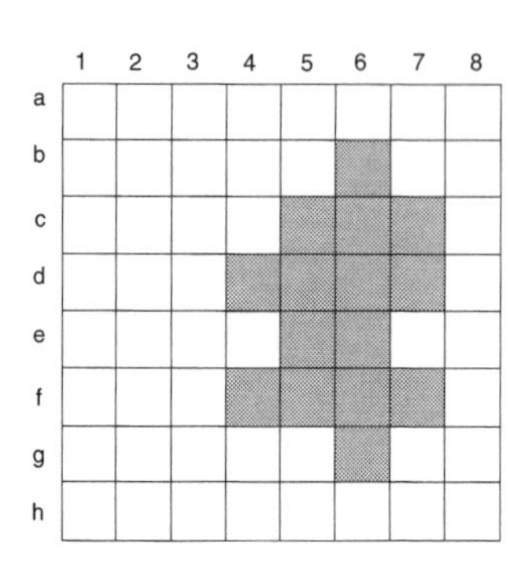

SCALE

The ratio or fraction between the distance on a map, chart, or photograph and the corresponding distance in the real world.

SCALE BAR

A map element which shows the scale graphically.

SCANNER

An instrument for data capture which converts an analogue image or map into digital raster form by systematic line-by-line sampling. Data is stored as an image with values representing the colour or grey tone detected at each corresponding location on the original.

See also FLATBED SCANNER *and* DRUM SCANNER.

SDSS

SPATIAL DECISION SUPPORT SYSTEM

A computer system designed to assist in the analysis of complex, ill-structured spatial problems. It is usually developed to aid strategic spatial planning decisions of organizations.

SDTS

SPATIAL DATA TRANSFER STANDARD

The US Federal Information Processing

Standard for transfer of spatial data (FIPS Publication 173).
See also NTF.

SEED

A point within a polygon that can be used to carry the attributes of the whole area, for example, ownership, address, land use type. This is often the CENTROID of the polygon.

SEMI-AUTOMATED DIGITIZING

A method of digitizing from a map in which the majority of the line following is controlled by software, but which requires an operator to be on hand constantly to assist in the identification of features and to resolve anomalies.

SHADED RELIEF MAP

An elevation map to which the shadow effects of natural light illumination are added.

SHORTEST PATH SOLUTION

See OPTIMAL PATH SELECTION.

SIEVE MAPPING

A term applied to the manual form of OVERLAY analysis. Maps drawn on transparent film are superimposed onto each other to ascertain coincidence and interaction.

SIF

STANDARD INTERCHANGE FORMAT

A proprietary format primarily used for the transfer of CAD drawings.

SINK

See PIT.

SLIVER POLYGON

1) A small area formed when two polygons which have been overlaid do not abut exactly along one or more edges.
2) A small area created during the digitizing or scanning of data.

Sliver polygons.

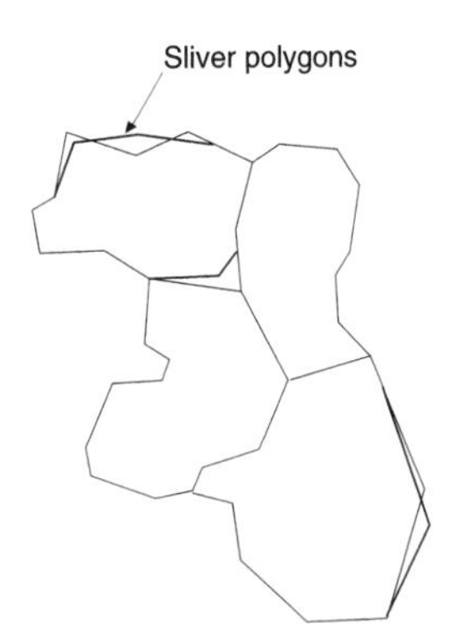

SNAPPING

An automatic editing operation in which points which fall near to other points or lines will be moved slightly so that they correspond.

SOFT COPY

A temporary image of a map or diagram on a screen or similar device.

SOFTWARE

The programs, procedures, rules, and their associated documentation, for a computer system.

SOFTWARE ENGINEERING

The systematic application of scientific and technological knowledge, methods, and experience to the design, implementation, testing, and documentation of software to optimize its production, support, and quality.

SPAGHETTI DATA

Vector data composed of line segments which are not structured in terms of TOPOLOGY or organized into spatial elements and which may not even be geometrically clean.

SPATIAL ANALYSIS

Analytical techniques whose results depend upon the location of the ENTITIES being studied. It includes the study of locations, dimensions, and ATTRIBUTE VALUES of geographic phenomena.

SPATIAL CORRELATION

A measure of the degree to which relative location determines the strength of relationships between different spatial elements.

SPATIAL DATA

Any type of data which includes a formal locational reference such as a grid reference. This includes remotely sensed as well as map data.

See also GEOGRAPHIC DATA.

SPATIAL DATA TRANSFER STANDARD

See SDTS.

SPATIAL DECISION SUPPORT SYSTEM

See SDSS.

SPATIAL QUERY

A query which selects features based on their location or geographic relationship to others.

SPATIAL REFERENCE

A co-ordinate, textual description or codified name by which an ENTITY can be related to a specific position or location in space.

SPATIAL RESOLUTION

A measure of the smallest area identifiable on an image as a discrete separate unit. In raster data, it is often expressed as the size of the raster cell. In remote sensing, it is defined in terms of the diameter of the ground area that be may be distinguished and is often comparable to the size of the earth's surface covered by a single pixel. The remote sensing systems on board the METEOSAT series of satellites have a spatial resolution of approximately 900 m whilst those on the SPOT series can resolve features down to 10 m in size.

SPECTRAL

An term to describe a defined range of wavelengths of ELECTROMAGNETIC ENERGY.

SPECTRAL REFLECTANCE CURVES

A generalized pattern of the absorption and reflection characteristics by a substance of ELECTROMAGNETIC ENERGY across a series of wavelengths.

Spectral reflectance curves of vegetation and water across the visible and infrared wavelengths of the electromagnetic spectrum.

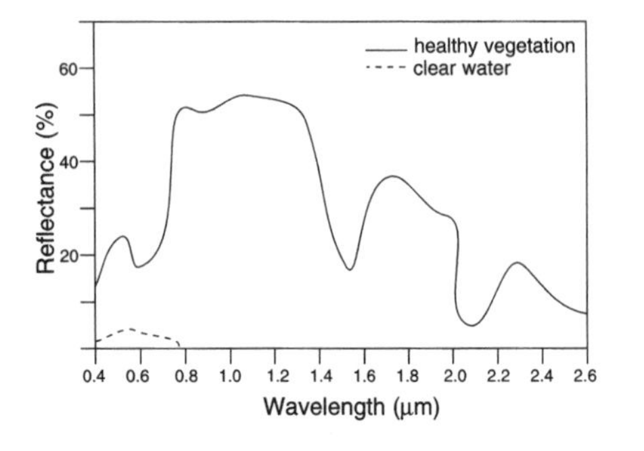

SPECTRAL RESOLUTION

The smallest waveband of ELECTROMAGNETIC ENERGY that is detectable by a sensor. It is usually defined in terms of micrometers. In meteorological satellite sensors the spectral resolution is relatively low (i.e. wide wavebands). In imaging spectrometry it is high (i.e. narrow wavebands) because the narrow wavebands are important in detecting the subtle differences in reflection and absorption patterns.

SPECTRAL SIGNATURE

See SPECTRAL REFLECTANCE CURVES.

SPECULAR REFLECTION

The idealised concept of the mirror-like reflection of ELECTROMAGNETIC ENERGY by a surface at an angle which is equal to the angle of incidence on the earth's surface.

Specular reflection of electromagnetic energy by the ground surface.

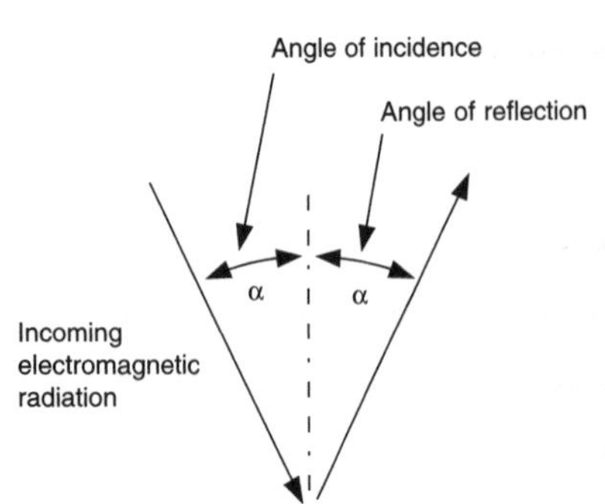

SPIKE

1) A protusion in the digital surface of a DTM. **2)** An OVERSHOOT created by mistake during the scanning or digitizing processes.

SPLINE

A smooth curve fitted mathematically to a sequence of points.

SPOT

SYSTÈME PROBATOIRE DE L'OBSERVATION DE LA TERRE A series of French polar-orbiting satellites which collect data suitable for Earth surface mapping and resource management.

SQL

STRUCTURED QUERY LANGUAGE

A standard language used with relational database products which enables the user to build complex logical expressions identifying the data to be recalled and manipulated.

STANDARD INTERCHANGE FORMAT

See SIF.

STEP

A transfer format for graphics data being developed by the ISO (TC184/SC4) to replace IGES.

STEREOSCOPE

A binocular optical device for viewing two images which have a slightly different view of the same scene. The person using the device observes a mental three-dimensional view of the scene.

STREAM DIGITIZING

This is a mode of DIGITIZER operation where points are recorded automatically at pre-set intervals of either distance or time.

STRING

A sequence of line segments or text items. It does not have topological properties.

STRUCTURED QUERY LANGUAGE

See SQL.

SUN-SYNCHRONOUS SATELLITE

See POLAR-ORBITING SATELLITE.

SURVEYING

The measurement and recording of geographically distributed information. Particular types of surveying are topographical, cadastral, and geological.

SYMBOL

A graphic representation of a concept that has meaning in a specific context. It is used in cartography to show the presence of ENTITIES such as churches, post offices, and public houses.

TABLE

A means of organizing data in rows and columns in which each row represents an individual entity, record, or feature and each column represents a single FIELD or attribute value.

TABLET

See GRAPHICS TABLET and DIGITIZER.

TAG IMAGE FILE FORMAT

See TIFF.

TCP/IP

TRANSMISSION CONTROL PROTOCOL/INTERNET PROTOCOL

This is a protocol for allowing computer systems to communicate and transfer data across the Internet. May be used for internal or external communications.

TESSELLATION

A subdivision of a 2D plane or 3D volume into discrete, contiguous spatial elements which completely cover the plane or volume. In GIS work, the terms regular tessellation and irregular tessellation are used. Regular tessellations use elements that are the same size and shape. A raster is an example of a regular tessellation. Irregular tessellations use a set of connected polygons of varied shapes and sizes which together completely cover an area. A TIN is an example of an irregular tessellation.

Regular and irregular tessellations of square, triangle, and hexagon type shapes.

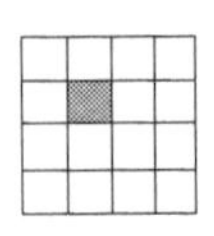
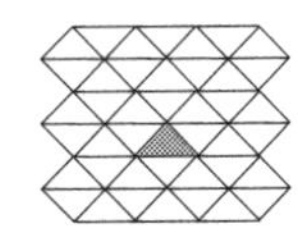
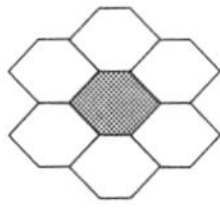
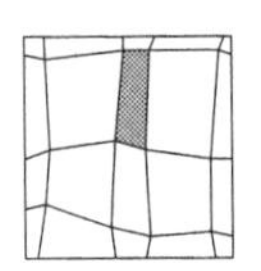
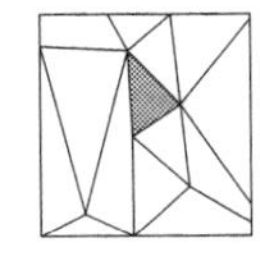
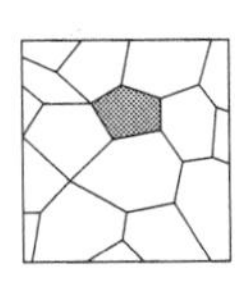

THEMATIC MAP

A map depicting one or more specific topics or subjects. Examples include land classification, population density, and rainfall maps.

THEMATIC MAPPER

TM A sensor on board the later LANDSAT satellite series. It detects radiation in seven wavebands and has a SPATIAL RESOLUTION of 30 m. The resulting data has provided important spatial information for earth scientists and resource managers.

THIESSEN POLYGONS

A method used to divide an area into polygons so that all locations closest to a particular sample point are enclosed within a single polygon. The boundary lines are defined at positions equidistant between two adjacent points. Also known as Dirichlet tessellations and Voronoi polygons.

THINNING

See LINE THINNING.

THRESHOLDING

A technique of data transformation which compares the values of an input data set to a specific given value, the threshold. A new data set will be constructed in which input values are converted to one of three output values, indicating whether the input value was over, equal to, or below the threshold.

TIFF

TAG IMAGE FILE FORMAT

A commonly-used raster file format developed by the Aldus Corporation.

TIGER

TOPOLOGICALLY INTEGRATED GEOGRAPHIC ENCODING REFERENCING

A spatial data format developed by the US Bureau of Census for the 1990 US census of population.

TIGHTLY COUPLED MODELLING

A term which refers to the linking of a GIS to a mathematical model through common data elements and certain system operations. Some processing and analysis of the data may be undertaken within the joined system.

TILE

A logical set of data covering a rectangular area which subdivides a large digital map data set into manageable units.

TIN

TRIANGULATED IRREGULAR NETWORK

A form of irregular TESSELLATION based on triangles and used to represent continuous spatial data originating as a set of irregularly spaced points. Unlike a grid, the TIN allows dense information in complex areas, and sparse information in simpler or more homogeneous areas. A TIN is often used to represent continuous elevation surfaces.

A triangulated irregular network (TIN) derived from a point data set.

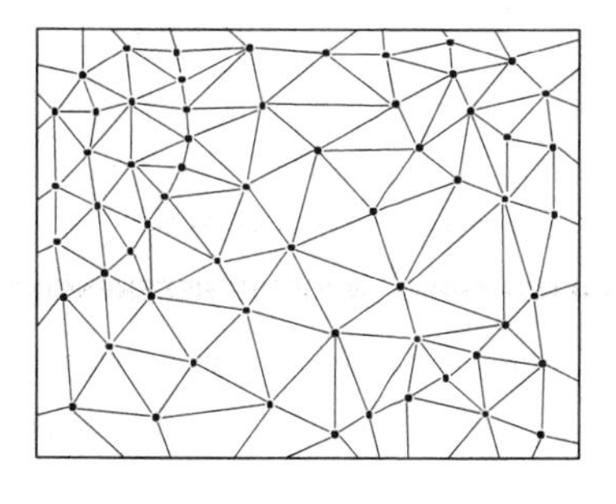

TINT

A stipple dot pattern used to create subdued colour infill within a defined area.

TM

See THEMATIC MAPPER.

TOLERANCE

A small distance, normally user-defined, which expresses the uncertainty with which a digitized position has been recorded. The term is often used to determine which OVERSHOOTS, UNDERSHOOTS, and redundant points should be corrected by SNAPPING.

TOPOGRAPHICAL DATABASE

An organized collection of data relating to the physical features and boundaries on the Earth's surface. Contains data formerly represented on topological maps.

TOPOLOGICALLY INTEGRATED GEOGRAPHIC ENCODING REFERENCING

See TIGER.

TOPOLOGICALLY STRUCTURED DATA

Data which is structured in such a way that relations and characteristics referred to as TOPOLOGY can be expressed. This would include concepts such as CONNECTIVITY, ADJACENCY and containment. For example, a line might have attributes including start and end nodes, direction, and identifiers for the areas to the left and right.

TOPOLOGICAL PROPERTIES

Those properties which define relative relationships between spatial elements in a database. These include ADJACENCY, CONNECTIVITY and containment but do not include angles and distance. Thus the relative location of geographic phenomena is independent of their exact position. For example, maps of metro or train networks usually show relative rather than actual location.

TOPOLOGY

Strictly speaking, the study of those properties of geometrical figures that are invariant under continuous deformation. In GIS, topological relationships, such as CONNECTIVITY, ADJACENCY and relative position, are usually expressed as relationships between nodes, links, and polygons.

TRANSFER FORMAT

A format for transferring digital data between computer systems and software. In general usage this can refer not only to the organization of data, but also to the associated information, such as attribute codes, which is required in order to complete the transfer successfully.

TRANSFER MEDIUM

The physical magnetic medium on which digital data is conveyed from one computer system to another. For example, magnetic tapes and floppy disks.

TRANSMISSION

In remote sensing the term refers to the passing of ELECTROMAGNETIC ENERGY through atmospheric or surface material.

TRANSMISSION CONTROL PROTOCOL/INTERNET PROTOCOL

See TCP/IP.

TRANSVERSE MERCATOR

See MERCATOR PROJECTION.

TRIANGULATED IRREGULAR NETWORK

See TIN.

TUPLE

A set of related values of attributes pertaining to a given item in a relational database. It is analogous to a RECORD in a TABLE.

TURN-KEY SYSTEM

A hardware and software system that is designed, supplied, and supported by a single commercial organization. It is ready for immediate use.

TWO-DIMENSIONAL DATA

See 2D.

UNDERSHOOT

A line which is short of its true intersection with another.

An undershoot.

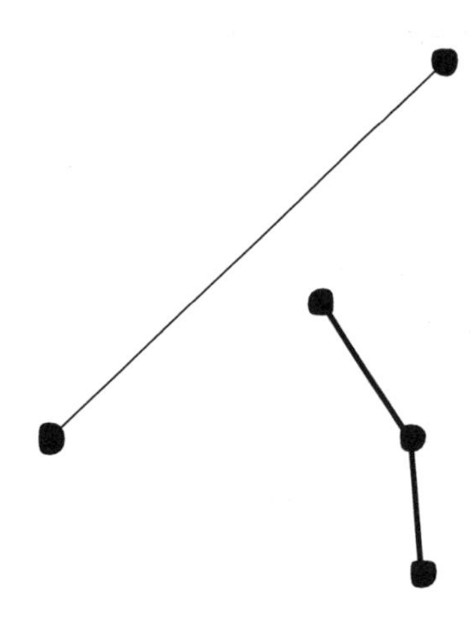

UNIQUE PROPERTY REFERENCE NUMBER

UPRN A short-coded address or number which uniquely identifies a land parcel.

UNIVERSAL TRANSVERSE MERCATOR

See UTM.

UPDATE

1) The process of adding to or revising existing data to take account of change.
2) A revision to a software product offering improved functionality and bug fixes.

UPRN

See UNIQUE PROPERTY REFERENCE NUMBER.

URBAN AND REGIONAL INFORMATION SYSTEMS ASSOCIATION

See URISA.

URISA

URBAN AND REGIONAL INFORMATION SYSTEMS ASSOCIATION

A US-based American and Canadian professional and educational organization to promote the effective use of information systems by local, regional, and state (in the US) or provincial (in Canada) government.

UTM

UNIVERSAL TRANSVERSE MERCATOR

A set of transverse MERCATOR PROJECTIONS for the globe which are divided into 60 zones, each covering 6 degrees longitude. The origin in each zone has the longitude of the central meridian and the latitude of 0 degrees. This set of projections is the basis of a global CO-ORDINATE SYSTEM developed initially for military applications but now widely used.

VECTOR DATA MODEL

An ABSTRACTION of the real world in which spatial elements are represented in the form of points, lines, and polygons. These are geographically referenced to a CO-ORDINATE SYSTEM.

Points, lines, and polygons of the vector data model.

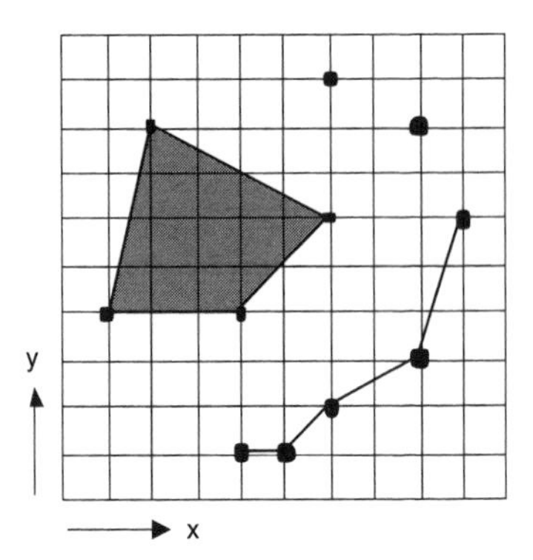

VECTOR-TO-RASTER CONVERSION

A process which transforms SPATIAL DATA held as a series of points, lines, and polygons into an ARRAY of cell values.

VIEWSHED

The boundaries of the area that would be seen from a particular vantage point, given the nature of the terrain and assuming an Earth surface without obstructions like trees and buildings. In a GIS this may be calculated and the results given as a map of the viewable area.

VORONOI POLYGONS

See THIESSEN POLYGONS.

VOXEL

A spatial element that represents 3D data in a regular TESSELLATION. The term is an abbreviation for 'volume element' and is analogous to a PIXEL in two dimensions.

WAVEBAND

A term used extensively in remote sensing to describe a contiguous range of wavelengths of electromagnetic energy.

WEEDING

See LINE THINNING (1).

WEIGHTING

The assignment of a significance value to a spatial element or class relative to its importance in a particular analysis. This value will determine the influence this element or class has on the final result.

WHAT-YOU-SEE-IS-WHAT-YOU-GET

See WYSIWYG.

WHISKBROOM SCANNING

See ACROSS-TRACK SCANNING.

WINDOW

1) A part of a display image with defined boundaries in which information, images, or commands are displayed and through which the user is able to interface with the computer. In many computer systems it is possible to have several windows open at the same time in order to see different views of the data or to control different parts of the program simultaneously.

2) A rectangular set of cells which is 'placed' over each cell in turn in order to determine which neighbouring cells will be considered during a FILTER operation. Such windows are generally 3 by 3 or 5 by 5 cells in size.

A window (3 by 3) of grid cells used in calculations.

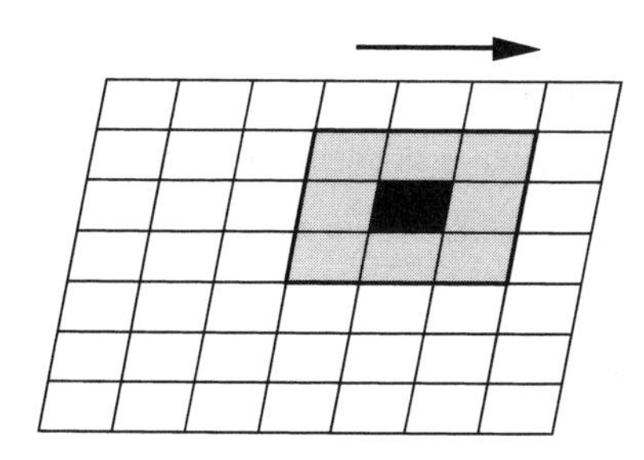

WIREFRAME REPRESENTATION

A graphical representation of a 2.5D or 3D solid or surface, composed entirely of regularly spaced lines, depicted as though the surface is constructed of wire. In GIS applications the lines often represent edges or surface contours in the display.

WYSIWYG

WHAT-YOU-SEE-IS-WHAT-YOU-GET

A capability to display information on a screen exactly as it will be printed or plotted on an output device.

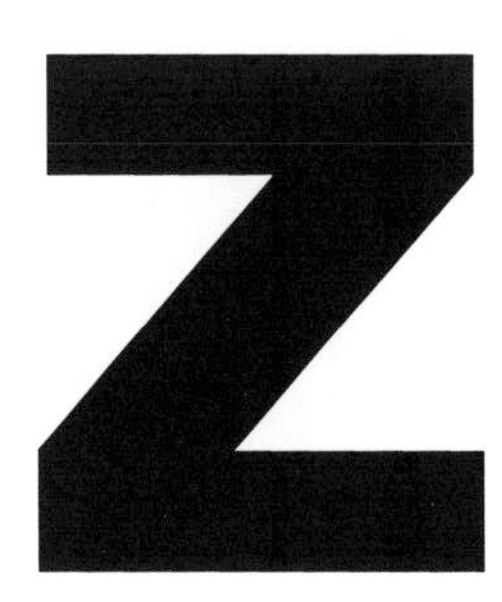

ZENITHAL PROJECTION

See AZIMUTHAL PROJECTION.

ZIPCODE

The US equivalent of the POSTCODE. It consists of five numbers plus an optional four (known as zip+4).

ZOOM

The capability for progressive scaling of the entire display image to give the visual impression of movement of display elements towards or away from the observer. A zoom operation changes the scale at which an image is being displayed.

Z-VALUE

A commonly used reference to ATTRIBUTE VALUES. Elevation is a commonly-used example of z-value.

ACRONYMS

AAG	Association of American Geographers
ABS	Australian Bureau of Statistics
ACORN	A Classification of Residential Neighbourhoods
ACSM	American Congress on Surveying and Mapping
ADMATCH	Address Matching Software
ADT	Abstract Data Type
AGI	Association for Geographic Information
AI	Artificial Intelligence
AIS	Address Information System
AIS	Australia Information System
AIT	Asian Institute of Technology
ALIC	Australian Land Information Council
AMEDAS	Automated Meteorological Data Acquisition System
AM/FM	Automated Mapping/Facilities Management
AML	ARC Macro Language
AMTICS	Advanced Mobile Traffic Information and Communication System
ANSI	American National Standards Institute
ARIS	Australia Resources Information System
ASCII	American Standard Code for Information Interchange
ASPRS	American Society for Photogrammetry and Remote Sensing
ATM	Adaptive Triangular Mesh
AURISA	Australasian Urban and Regional Information Systems Association
AUSLIG	Australian Surveying and Land Information Group
AVHRR	Advanced Very High Resolution Radiometer
AVIRIS	Airborne Visible-Infrared Imaging Spectrometer

BIT	Binary Digit
BLM	Bureau of Land Management (US)
BLOB	Binary Large Object
BLPU	Basic Land and Property Unit
BLUE	Best Linear Unbiased Estimate
BSI	British Standards Institute
BSU	Basic Spatial Unit
BURISA	British Urban and Regional Information Systems Association

CAD	Computer-Aided Design
CAD	Computer-Aided Drafting
CADD	Computer-Aided Design and Drafting
CAG	Canadian Association of Geographers
CAL	Computer-Assisted Learning
CAM	Computer-Aided Mapping
CARD	Cartographic Representation of Data
CARP	Computer-Assisted Regional Planning
CASS	Crime Analysis Statistical System
CCD	Charge-Coupled Device
CCITT	Comité Consultatif International de Télégraphique et Téléphonique
CCT	Computer-Compatible Tape
CD-ROM	Compact Disk Read Only Memory
CEN	Comité Européen de Normalisation

CERCO	Comité Européen des Responsables de la Cartographie Officielle
CERES	Clouds' and Earth's Radiant Energy System
CGIA	Center for Geographic Information and Analysis
CGIS	Canada Geographic Information System
CGM	Computer Graphics Metafile
CHEST	Combined Higher Education Software Team (UK)
CIG	Canadian Institute of Geomatics
CISM	Canadian Institute of Surveying and Mapping
CLDS	Canada Land Data System
CLI	Canada Land Inventory
CLISG	Commonwealth Land Information Support Group
CNES	Centre National d'Etudes Spatiales
CNIG	Centro Nacional de Informação Geográfica (Portugal)
CNIG	Centre National d'Information Géographique (France)
COGO	Co-ordinate Geography
CORINE	Co-ordinated Information on the European Environment
CPGIS	Association of Chinese Professionals in Geographic Information Systems–Abroad
CRIES	Comprehensive Resource Inventory and Evaluation System
CRISP	Computerized Rural Information Systems Project
CSIRO	Commonwealth Scientific and Industrial Research Organisation (Australia)
CZCS	Coastal Zone Color Scanning

DBMS	Database Management System
DCDSTF	Digital Cartographic Data Standards Task Force
DCM	Digital Cartographic Model
DDE	Dynamic Data Exchange

DEM Digital Elevation Model
DFUS Digital Field Update System
DIGEST Digital Geographic Information Working Group Exchange Standard
DIME Dual Independent Map Encoding
DIP Document Image Processing
DLG Digital Line Graph
DLG-E Digital Line Graph – Enhanced
DLM Digital Landscape Model
DMA Defense Mapping Agency (US)
DMSP Defense Meteorological Satellite Program
DN Digital Number
DPI Dots Per Inch
DRIVE Dedicated Road Infrastructure for Vehicle Safety
DSS Decision Support System
DTED Digital Terrain Elevation Data
DTM Digital Terrain Model
DXF Digital Exchange Format

EC European Commission
ECU Experimental Cartography Unit
ED Enumeration District
EDI Electronic Data Interchange
EOS Earth Observation Satellite
EOSDIS EOS Data Information System
EPA Environmental Protection Agency (US)
ERIN Environmental Resources Information Network
ERS Earth Resources Satellite of the European Space Agency and also of Japan
ES Expert System

ESA	European Space Agency
ESF	European Science foundation
ESRC	Economic and Social Research Council
ESRI	Environmental Systems Research Institute
EUROGI	European Umbrella Organization for Geographic Information

FGDC	Federal Geographic Data Committee (US)
FICCDC	Federal Interagency Co-ordinating Committee on Digital Cartography (US)
FIG	Fédération International de Géomètres
FIPS	Federal Information Processing Standard (US)
FTP	File Transfer Protocol

GAM	Geographical Analysis Machine
GB	Gigabyte
GBF-DIME FILE	Geographic Base File
GCDP	Global Change Database Project
GCP	Ground Control Point
GDF	Geographic Data File
GIA	Geographic Information Analysis
GIRAS	Geographic Information Retrieval and Analysis System
GIS	Geographic Information System
GISA	Geographic Information Systems Association
GISG	Geographic Information Steering Group

GKS	Graphical Kernel System
GOES E AND W	Geostationary Operational Environmental Satellites East and West
GPS	Global Positioning Satellite
GPS	Global Positioning System
GRASS	Geographical Resources Analysis Support System
	Global Resource Information Database
GSI	Geographic Survey Institute
GSM	General Systems Model
GSS	Government Statistical Service
GUI	Graphical User Interface

HCI	Human-Computer Interaction
HCI	Human-Computer Interface
HCMM	Heat Capacity Mapping Mission
HDGCP	Human Dimensions of Global Change Program
HMSO	Her Majesty's Stationary Office

ICA	International Cartographic Association
ICSU	International Council of Scientific Unions
IEEE	Institute of Electrical and Electronics Engineering Inc.
IGBP	International Geosphere-Biosphere Programme
IGES	International Graphics Exchange System
IGGI	Inter-Departmental Group on Geographic Information
IGU	International Geographic Union

IHO	International Hydrographic Organization
IJGIS	International Journal of Geographical Information Systems
ILWIS	Integrated Land and Watershed Management Information System
IMM	Interactive Multi-Media
IMS	Information Management System
IOC	International Oceanographic Commission
IODE	International Oceanographic Data and Information Exchange
ISDN	Integrated Services Digital Network
ISM	Interactive Surface Modelling
ISO	International Standards Organisation
ISPRS	International Society of Photogrammetry and Remote Sensing
ISSS	International Soil Science Society
ISTF	Intermediate Standard Transfer Format
ITC	International Institute for Aerospace Survey and Earth Science
ITE	Institute of Terrestrial Ecology (UK)
IUGG	International Union of Geodesy and Geophysics

J

JANET	Joint Academic Network (UK)
JDRMA	Japan Digital Road Map Association

KB	Kilobyte
KBGIS	Knowledge Based Geographic Information System

LAMIS	Local Authority Management Information System
LAN	Local Area Network
LAS	Land Analysis System
LCG	(Harvard) Laboratory for Computer Graphics
LIS	Land Information System
LOTS	Land Ownership and Tenure System
LREIS	Laboratory for Resource and Environmental Information Systems
LRIS	Land Resources Information System

MAP	Map Analysis Package
MB	Megabyte
MDS	Multi-Dimensional Scaling
MEGRIN	Multi-Purpose European Ground Related Information Network
MIS	Management Information System
MMU	Minimum Mapping Unit
MOSS	Map Overlay and Statistics System
MSS	Multi-Spectral Scanner

NA	Network Analysis
NAG	Numerical Algorithm Group
NASA	National Aeronautics and Space Administration (US)
NCDC	National Climate Data Center (US)
NCGA	National Computer Graphics Association
NCGIA	National Center for Geographic Information and Analysis (US)
NCIC	National Cartographic Information Center
NDCDB	National Digital Cartographic Database
NDVI	Normalized Difference Vegetation Index
NERC	Natural Environmental Research Council (US)
NEXPRI	Nederlands Expertise Centruum voor Ruimtelijke Informatiererwerkig
NGDC	National Geophysical Data Center (US)
NIR	Near Infrared
NJUG	National Joint Utilities Group
NOAA	National Oceanographic and Atmospheric Administration (US)
NODC	National Oceanographic Data Centres
NOMIS	National On-Line Manpower Information System
NRIC	National Resources Information Centre
NSDI	National Spatial Data Infrastructure
NSF	National Science Foundation (US)
NSSDC	National Space Science Data Center
NTF	National Transfer Format

OGIS	Open Geodata Interoperability Specification
ONC	Operational Navigation Charts
OODB	Object-Oriented Database
OODBMS	Object-Oriented Database Management System
OPCS	Office of Population Census and Surveys (UK)
OS	Ordnance Survey (UK)
OSNI	Ordnance Survey of Northern Ireland
OSTF	Ordnance Survey Transfer Format (UK)
OSTF+	Ordnance Survey Transfer Format Plus (UK)

PANDORA	Prototyping a Navigation Database of Road Network Attributes
PDES	Product Data Exchange Specification
PHIGS	Programmer Hierarchical Interactive Graphics System
PMAP	Professional Map Analysis Package
PUSWA	Public Utilities Street Works Act

RADAR	Radio Detection and Ranging
RDBMS	Relational Database Management System
REGIS	Regional Geographic Information Systems Project
RGS	Royal Geographical Society (UK)
RIN	Royal Institute of Navigation

RLE	Run-Length Encoding
RLUIS	Rural Land Use Information System
RNODC	Responsible National Oceanographic Data Centres
ROADIC	Road Administration Information Centre
RRL	Regional Research Laboratory (UK)
RTPI	Royal Town Planning Institute (UK)

SACS	Small Area Census Studies
SAR	Synthetic Aperture Radar
SASPRSC	South African Society for Photogrammetry, Remote Sensing, and Cartography
SDSS	Spatial Decision Support System
SDTS	Spatial Data Transfer Standard (US)
SET	Système d'Échange et de Transfer
SIC	Standard Industrial Classification
SIF	Standard Interchange Format
SIS	Spatial Information System
SLDS	Swedish Land Databank System
SORSA	Spatially Oriented Referencing Systems Association
SPOT	Système Probatoire de l'Observation de la Terre
SQL	Structured Query Language
STI	Scientific and Technical Information
SUSI	Sale of Unpublished Survey Information

TB	Terabyte
TCP/IP	Transmission Control Protocol/Internet Protocol
TIFF	Tag Image File Format
TIGER	Topologically Integrated Geographic Encoding Referencing (US)
TIN	Triangulated Irregular Network
TM	Thematic Mapper

UIS	Urban Information System
UNDP	United Nations Development Programme
UNEP	United Nations Environmental Programme
UNITAR	United Nations Institute for Training and Research
URISA	(North American) Urban and Regional Information Systems Association
URPIS	Urban and Regional Planning Information Systems
USBC	United States Bureau of the Census
USDA	United States Department of Agriculture
USGS	United States Geological Survey
UTM	Universal Transverse Mercator

VNIS Vehicle Navigation and Information Systems Conference

WAN Wide Area Network

WMO World Meteorological Organization

BIBLIOGRAPHY

American Cartographer 15 (1) 1988

Aronoff S 1989 *Geographic Information Systems: A Management Perspective.* Ottawa, WDL Publications

Burrough P A 1986 *Principles of Geographical Information Systems for Land Resources Assessment.* Oxford, Clarendon Press

Lillesand T M, Kiever R W 1994 *Remote Sensing and Image Interpretation* 3rd edition. New York, Wiley & Sons, Inc

Maguire D, Goodchild M F and Rhind D 1991 *Geographical Information Systems: Principles and Applications.* Harlow, Longman Group Ltd

Glossary of Terms Technical Report 92 (13) 1992. Santa Barbara, NCGIA

1994 *International GIS Sourcebook.* Fort Collins, GIS World, Inc